tech talk 2021

A COLLECTION OF THE BEST TECH TALK COLUMNS

Tech Talk 2021

by

Mathew Dickerson

A collection of the
best Tech Talk columns

TECH TALK 2021

A COLLECTION OF THE BEST TECH TALK COLUMNS

ISBN: 9780648782711 (978-0-6487827-1-1)
Word Count: 51699
Flesch Reading Ease: 55.6
Flesch-Kincaid Grade Level: 9.9

Publisher: Small Business Ru!es
9 8 7 6 5 4 3 2 1 0

The author can be contacted at ask@techtalk.digital

SMALL BUSINESS RU!ES

DEDICATION

In writing this dedication to my readers, I really just want to say THANK YOU! When I sit down to write my weekly column, I try and imagine some of my readers reading my column. I hear feedback all the time and the variety of the feedback astounds me. Some readers have a casual read and enjoy the entertainment value with a vague interest in the technology. Other readers drift in and out of reading my columns depending on their interest level on specific topics. Other readers tell me they pore over every detail and pride themselves on being the technical expert at their next BBQ or night at the pub.

With a huge variety of readers, it puts pressure on me to perform each week but also inspires me to make every column a masterpiece.

I start off with an idea. It might be something in the news. There may have been a conversation with a client that sparked my curiosity about a topic. I am curious by nature and if I have questions then maybe other people have similar questions.

Once I decide on a topic, I research it. And research it. And research it. If my readers are quoting information from my columns I need to make sure I have the data correct.

Once I have the topic idea and the relevant data, I then like to think of an interesting introduction that relates to the topic but doesn't dive in immediately. Some personal experience or some historical moment.

Then I write and write. Once I start on a topic, I normally take a breath at over 900 words. But my word limit is 600. Then it becomes challenging. Mark Twain said once that he wrote a long book because he didn't have time to write a short book!

Mark Twain probably had the luxury of being somewhat flexible with his words but my editor needs to fit my column in to a certain space, so each column is precisely 600 words. Count them if you must but that seems just the right amount to explore a topic but keep readers interested.

I hope you enjoy reading these as much as I enjoy writing them.

Mathew

Contents

FOREWORD

A collection of opinion columns about all things technological, Tech Talk is the thought-provoking, yet unassumingly humorous, opinions of technologist and futurist Mathew Dickerson.

Not your typical jargon-heavy tech column that leaves most with eyes-a-glaze, Dickerson injects anecdotal humour, fun facts and the odd pun, to balance the techy speak for a more digestible read that offers insight and intrigue for those who don't run with the 'tech head' scene.

Teaming with interesting information and laced with just the right amount of tongue-in-cheek, Tech Talk is a valuable contribution to more than 150 regional newspapers across the Australian Community Media network. Through it, Dickerson brings a voice to topics most didn't know, they didn't know - a chance to keep up with the fast-moving world of futuristic bells and whistles.

But within the fun and humour, just as importantly, Dickerson brings inquisitivity with his opinions. I, like many of our readers, have been left pondering his insights, and even a little inspired by an industry that can sometimes seem out of reach in everyday life.

But that's where Dickerson bridges the far-off tech world with the front door of the family home, covering topics that are current and relevant.

Articles about artificial intelligence helping around the home with something as simple as doggie doors you can control from your smartphone, smart toilets that could change the way we keep our health on track and apps than can smell if the use-by date on our food is accurate, all play their part to engage readers with a different point of view of the tech world - and the idea that it's not as far away as we think.

Tech Talk opens up the world of technology as being more than just out-of-reach futuristic ideals and complicated gadgets - it's a great read with the added bonus of maybe even teaching us something. Happy reading.

Jess Wallace
Contributed Content Journalist at Australian Community Media

Introduction

RING! RING! *(In deep voice)* "This is Mathew Dickerson calling from the future...here to make technology accessible to you all!"

Now...despite your age, experience or confidence, I have three questions for you. Does technology scare you? Do you think technology will tear us apart? And how far do you believe someone will dare to go?

Scare...tear...dare...

Will you dare to tear this book open full of technological developments and scare the change that will occur in society?

In writing this book, I acknowledge that the majority of people have a good understanding of technology, however, my aim is to make technology accessible to everyone. I understand the worries of technology and the power it holds, but technology is best when it brings people together. I often use anecdotes and punny puns in my stories to make it them interesting!

As an added challenge, I have created a title for each technology column that adheres to a self-imposed rule: it must consist of three words that rhyme, encapsulate the essence of the chapter's content, and strike a balance between clarity, creativity, and thematic relevance. This naming convention aims to create a rhythmical flow, drawing readers in while offering a sometimes-cryptic preview of each chapter's focus.

For example, to describe the entire book, I could use: "Gear Sphere Here"

The book covers many technological tools (gear) that are used across the entire planet (sphere) and they are relevant to readers now or in the near future (here). Please keep an open mind as I have used a touch of poetic license with some.

So...rip open the world of technology and explore the cunning ideas and concepts our world is capable of.

Peer Year Cheer

This Christmas, I promised myself, I definitely wouldn't overindulge and then collapse in front of the TV and start pumping out zeds. But with a beautiful meal prepared and family gathered, the inevitable happened and before I knew it, I looked like Homer Simpson on the couch complete with dribble running from the corner of my mouth.

In much the same way I promised myself this year that I wouldn't do a 'Year in Review' column but the inexorable draw of the end of year retrospective column has consumed me and, try as I may, I can't help but look back over some of the technology from this year.

For the record, this is NOT an end of decade review. In the same way that all those parties on 31 December 1999 to celebrate the end of the century got it wrong, the end of this decade doesn't occur until the end of 2020. Phew! I got that off my chest!

Despite the unfounded health concerns, the 5G network is one of my highlights. Although technically Telstra switched on 5G in ten cities last year, this year has seen an increased adoption of 5G with over twenty-five cities across the nation having access to 5G and, just as importantly, several devices available that support 5G, including a mobile broadband Wi-Fi device. Australia is one of only ten countries with 5G coverage of any note.

With promises of greater speed, more capacity and lower latency, 5G will enable the unlocking of a huge number of other technology advances including the world of IoT (Internet of Things).

The wearables market is the next area that saw some giant leaps in 2019. Who needs a doctor when you have a watch to assess and monitor your personal health? Watches now allow continuous heart rate monitoring; ECG; sleep tracking; cardio fitness measuring; menstrual cycle tracking and a great feature that I cannot convince my children to use. Noise monitoring to help the wearer of a watch determine when a concert may be damaging your hearing (you would think the ringing sound for two days following a concert would be a subtle hint!) With international emergency calling with fall detection, wearers really feel safer and healthier with their latest wearable.

With the health of the planet in mind, electric cars have been around for some time but this year added some interesting twists. Lotus introduced a fully electric supercar; Lamborghini produced a hybrid production car and the bastion of petrol-heads, Harley Davidson, added electric motors to their mix. The 2020 Olympics will see autonomous electric vehicles being used to transport athletes. And the darling of the EV industry, Tesla, introduced their Cybertruck - with mixed reviews.

There were some negatives as well. Though technically the Cambridge Analytica scandal was a 2018 news story, the fallout continued in 2019 with the Federal Trade Commission issuing a US$5 billion fine and privacy concerns given as the main reason for the loss of 15 million Facebook users.

Unfortunately, I can't have a year in review without mention of the Samsung Galaxy Fold, the revolutionary folding mobile phone that threw in other features for free such as delamination and fragility. Foldable phones will be in our future, just not part of 2019.

The Boeing 737 MAX also shows how we can sometimes rely on technology too much. Complete trust in an overaggressive automation system was given as a direct contributor to two crashes with significant loss of life.

While I prepare for some snoozing during the Boxing Day Test, tell me your biggest technology moments of 2019 at ask@techtalk.digital. Tech Talk 178: 27 Dec 2019

Hypothesise Materialise Editorialise

Data shows that 92 per cent of all resolutions made in the heady euphoria of 1 January are unsuccessful so I thought I may as well start the New Year by breaking my recurring resolution.

Every year I tell my readers it is simply too hard to make one-year technology predictions so I promise that each year will be my last...yet here I am again making predictions that I know are going to make me look silly in 12 months.

Ripping that band-aid right off...

Electric vehicles: Global sales of pure Electric Vehicles (EV) hit 1.5 million for 2019 and even though this represents only 2 per cent of global car sales, it is a figure that has been doubling each year for three years. EVs in Norway accounted for 40 per cent of all sales in 2019.

Prediction: The world will hit the 3 million mark for EV sales this year representing approximately 4 per cent of all sales. The EV industry itself has a policy position of EV30@30 which states a target of 30 per cent of all sales by 2030.

Streaming services: Netflix is currently king of the castle with 139 million subscribers. Amazon Prime sits hot on the heels with 100 million. Saudi Arabia has an 86 per penetration rate followed by 80 per cent in the US. Australia sits in seventh place at 70 per cent.

Prediction. As king, Netflix may say that Disney+ and Apple TV+ are the dirty rascals but as new kids on the block, they will impact both Netflix and Amazon. I expect to see these two new streaming services with a combined subscriber base of 60 million by the end of the year. Netflix and Amazon numbers will stagnate and Australia will reach a 75 per cent penetration rate.

Internet traffic: Global data volume in 2019 sat at 167 exabytes (EB) per month. Netflix makes up 12 per cent of this volume!

Prediction: By year end, we will see global data volumes go past 210 EB per month. This figure will include 42 EB of mobile data which will be somewhat driven by the penetration of 5G around the world.

Cybercrime: In 2019, cybercrime was a US$1.6 trillion industry and the fastest growing criminal activity in the world. It is a complete coincidence that I happen to be travelling through Spain at the moment but this country was the hardest hit by cyberattacks with 93.7 per cent of all companies compromised at least once last year.

Prediction: The increase in Internet traffic volumes and speed of connections will see an increase in cybercrime this year of 20 per cent making it a US$2 trillion industry. That ranks cybercrime higher than the GDP of Australia!

Areas to watch in 2020: We have a range of voice assistants in our houses and smartphones but the one area where we could really use some handsfree assistance – in our cars – has been somewhat lacking. Expect to see better voice assistance in cars this year. While on cars, expect to see cars on the road delivering on the autonomous driving promise by the end of the year. The wearable markets exploded last year but there is still more potential there. I believe we will see some incredible breakthroughs with wearables and biometric data this year. Lastly, I hope, more than predict, that this is the year that work will be finally something that you do rather than a place you go to and remote work starts to solve traffic congestion across the world.

Tell me your number one technology prediction for this year at ask@techtalk.digital. TECH TALK 179: 3 JAN 2020

Unorthodox Gearbox Rocks!

One of the traits I most admire about the human race is the ability to be creative and innovative in solving problems. Go back to circa 3500 B.C. when the Mesopotamians invented the wheel or the Egyptians invented the nail as a fastener then jump to 1439 when Johannes Gutenberg invented the printing press or consider more modern inventions such as light bulbs and telephones and vaccines and...well, name your favourite invention or discovery. The list is extensive.

It is obvious to all and sundry (apart from our government) that we need to come up with alternative fuel sources for transportation. Electric vehicles are making slow inroads (pun out of the way for this week) and even electric planes are being trialled for short flights.

An alternative fuel source that is often discussed is hydrogen. Some car companies are basing a large part of their future strategy on hydrogen rather than electricity but hydrogen also has the potential to power aircraft. Now when I mention hydrogen and flying the first image that comes to mind is the horrific image of the Hindenburg and the 36 people that died when the airship went up in flames.

One of the issues with using hydrogen is storage. Strong and large tanks are required that can withstand extremely high pressure. Usually in the order of 700 atmospheres. This creates a whole range of problems.

And then comes along human ingenuity.

Instead of storing the hydrogen as a gas under extreme pressure, could it be stored as a solid without the high levels of compression? Well one company thinks so. Trials are already underway with drones powered by solid state hydrogen. The hydrogen is mixed with oxygen in the reactor to produce power with a waste product of H_2O – good old-fashioned water.

While fuels based on oil have been the dominant fuel source for over a century it is almost surprising we haven't worked on hydrogen previously with the superior fuel capacity of hydrogen. While it is obviously the climate that is driving the change now, the numbers look good. Hydrogen can generate approximately 39kWh from one kilogram compared to 13kWh for petrol and only 0.2kWh for batteries. This superior fuel capacity by weight is particularly important in aircraft where every kilogram is important.

While the main focus of current development is the storage of hydrogen while waiting to be utilised, another major hurdle has already been largely addressed. Producing the initial hydrogen (typically by splitting water molecules) used a lot of energy which usually came from fossil fuels. This defeated the purpose of developing a clean energy source and also incurred significant costs. However, the improvements in electrolysis and the widespread availability of renewable energy have brought down the environmental and financial costs of producing hydrogen as a fuel source.

As solid-state hydrogen is further developed, I can see other uses for it as well. We know that large scale batteries lose charge over time while sitting idle waiting to be used. Solid state hydrogen has no such problem so it could be used for backup power storage facilities. Other transportation is also an obvious target market. Imagine a truck or train with the capability to travel three times the distance on the same weight of hydrogen fuel compared to the petrol equivalent.

There is no doubt in my mind that we have some significant technical challenges facing us in the immediate future – perhaps some of our biggest ever. But I have faith in human ingenuity to solve these problems for all mankind. Tell me what you believe man's greatest invention is at ask@techtalk.digital. TECH TALK 180: 10 JAN 2020

Televise Resize Prize

Henry Ellsworth, former Commissioner of the U.S. Patent Office, delivered his annual report to congress. In that report he stated: "The advancement of the arts, from year to year, taxes our credulity and seems to presage the arrival of that period when human improvement must end." The year? 1843. In the previous year, a record 490 patents had been approved. It is hard to say when human improvement will end but in 2019, 311,000 patents were approved by that same office.

One of the greatest challenges for developers is in a mature market. How do you continue to innovate and attract clients to the upgrade market when all possible innovation has seemingly occurred?

It is a dangerous thought pattern to fall into. Just when we thought the mobile phone market had reached a certain level of maturity, along came a complete disruption with the smartphone and the process started all over again.

> One product category that I constantly hear has reached a stalled state of maturity is the television.

From 30 June 1925 when Charles Jenkins was granted U.S. patent 1,544,156 for 'Transmitting Pictures over Wireless' to CRT and the flat-panel revolution of Plasma then LCD; LED; TFT and OLED, there is no doubt that the technology has progressed to the point when some question the ability to innovate further.

With the number of worldwide connected television sets predicted to pass the one-billion-mark next year and with annual sales approaching a quarter of a billion, market leading companies such as Samsung; LG; Sony and Hisense are literally banking on innovation to keep driving sales.

So where are we headed?

There are three main areas that the manufacturers are hoping will entice you to choose their version of your next purchase. Resolution, size and concealment.

4K resolutions (named for the number of pixels across the screen – 3,840 pixels wide by 2,160 pixels high for a total of 8.3 million individual pixels) would seem like the latest resolution but with 35 per cent of televisions sales last year at 4K the new battleground is already shaping up with 8K – a massive 33 million pixels on the screen. It seems hard to justify with only a few channels in Australia even broadcasting in 4K plus the limitation of the human eye – but someone forgot to tell the manufacturers as they will try and convince you that 8K is the only option to even consider now.

Remember when plasma TVs hit 42" and we thought that was a huge size for a TV? Before long, 42" became the standard fare for your new TV but that size is almost up to 65" now. In 2020 it is expected that 65" screens will make up 24 per cent of all sales and 75" screens – which were unheard of only a few years ago – will be up to 10 per cent. Bezels as small as 2.3mm make it even easier to fit the large screens in the same space – which brings me to my next point.

There was a time when you wanted your screen to stand out as a centrepiece of your living room. Manufacturers are betting that the opposite now applies. Screens that are 20mm thick to sit flush on your wall will be the basic choice. Televisions are starting to roll-up when not being used or, my favourite option, televisions that appear to be a piece of art hanging on the wall until a voice command springs them into life for your favourite viewing.

When will innovation end? I hope never. Tell me what you want in a TV at ask@techtalk.digital. TECH TALK 181: 17 JAN 2020

Handset Threat Onset

Put your hand up if you would like information on your mobile phone to be secure. Photos of your kids that you have taken on your phone to be kept private. E-mails from work about new product launches to remain just with employees. Posting information on social media to be limited to you. That type of thing.

I would be surprised if your hand is not currently in the air. There aren't too many people who would be happy if everything they have on their phone – their bank accounts, contacts, everything – ended up in the hands of a random person somewhere in the world.

Consequently, phone manufacturers place a high level of importance on security. Seems sensible.

Now jump to a separate conversation.

With an average of 1.2 mass shooting events per day in the US, you can be excused for not remembering one specific event. On 2 December 2015, 14 people were killed and 22 injured in an Islamic terrorist attack in San Bernardino. The two shooters were killed in a shootout with police but in investigating the case, the FBI found 400 pieces of evidence including an iPhone.

Exploring the prospect that the phone may have crucial evidence on it – including the likelihood that it may have information on a possible third shooter – the FBI politely asked Apple to unlock the phone of a known terrorist. Apple politely declined. By the end of 2016, Apple had received – and objected to – at least eleven orders issued by the United States district courts under the All-Writs Act of 1789.

I am not Apple but if the FBI knocked on my door and asked for some help in the investigation of a known terrorist, I would be pretty happy to bend over backwards. Most people would be the same – but in this case, go back to my original point. People want security in their phones. Apple argued that if they built a phone that had a secret backdoor that allowed them to access information then that same backdoor could be used by people with nefarious intent. Unlike Meghan and Harry, you can't have your cake and eat it too. As tragic as the loss of 14 lives was, Apple argued that the loss of lives would potentially be greater if Apple deliberately built a phone that was able to be hacked.

This case is ancient history in technology terms but the issue is once again at the centre of technology discussions after a lone shooter killed 3 and injured 8 in Pensacola, Florida on 6 December 2019. The incident has been officially classified as an act of terrorism.

Two iPhones have been recovered by the FBI as part of its investigation. In a case of history repeating itself, the US Attorney-General has urged the tech giant to unlock the two mobiles used by the gunman but, no surprises here, Apple have refused. They have explained that they "have always maintained there is no such thing as a backdoor just for the good guys," and further that "backdoors can also be exploited by those who threaten our national security and the data security of our customers."

It would seem that the logic from Apple is that the two things we desire in our society – security for our own mobiles but no security for mobiles of 'bad guys' – are mutually exclusive.

On the plus side for Apple, the ongoing vitriol from Trump and the US government is the best advertising possible for Apple's security!

Let me know where you stand on Apple's stance. Security for all or access for the government? ask@techtalk.digital. TECH TALK 182: 24 JAN 2020

Vision Imprecision Revision

"Gentlemen, we can rebuild him. We have the technology. We have the capability to make the world's first bionic man. Better than he was before. Better, stronger, faster." And so began each weekly episode of the 'Six Million Dollar Man' which ran from 1973 to 1978.

Steve Austin, the fictional bionic man, had his legs replaced to allow him to run at 60mph; one eye replaced to give him zoom level of 20.2 times and his right arm replaced with a unit that gave him the 'strength of a bulldozer'.

Very futuristic for a TV show from almost fifty years ago.

Take fiction and add lots of years of research and development and you can sometimes end up with reality. How are we progressing in our current environment with bionic replacements?

A world record was recently shattered at the Charleston Marathon by a man in an exoskeleton. Adam Gorlitsky crashed his vehicle fourteen years ago which left him paralysed below the waist. He purchased a robotic exoskeleton for $80,000 which allowed him to complete the marathon in 33 hours and 16 minutes. OK, maybe not better, stronger, faster but certainly a drastic improvement over no movement at all.

João Carlos Martins is an acclaimed classical pianist reaching the twilight of his career but a number of circumstances (including age) have made it increasingly difficult for him to play the piano. He now wears bionic gloves that help his hands maintain their function. As one of the greatest living interpreters of Bach, he is not yet at a standard that is acceptable to him but still hopes to play at Carnegie Hall in October to celebrate the 60th anniversary of his first appearance there.

Keven Walgamott lost his left hand and part of his arm in an electrical accident seventeen years ago. Thanks to a biomedical team at the University of Utah, he recently used his thoughts to not only move his prosthetic hand, but move it with enough feeling to pick up an egg without crushing it.

With an estimated 285 million people worldwide with visual impairment, development of a bionic eye would be a financial bonanza to any company that was successful. The best we have at the moment is a system that delivers light, movement and shapes at a low resolution with no colour. The main limitation is the fact that only 60 electrodes are used to transmit information from the outside world to the optic nerve. To see naturally would require about a million. At $150,000 the device is not priced for everyone but watch this space (I at least made it past the halfway mark before the weekly pun).

We have many people in society that have parts of their bodies replaced with various man-made items. From mechanical valves in hearts to complete artificial hearts. From cochlear implants to insulin pumps. We have become accustomed to various components being deployed in our bodies and, if it improves the quality of life for an individual, I believe that most people are comfortable with the concepts.

These examples are all aimed at restoring or repairing rather than improving the functionality of a human. Someone is paralysed or loses a hand or an arm. The issue becomes much more complicated when it relates to the Steve Austin scenario. Better, stronger, faster. Medical advancements are working very hard to use whatever means to make people better again but not so much stronger or faster. And that seems fair enough.

Tell me who was your favourite – Steve Austin, the Six Million Dollar Man or Jaime Sommers, the Bionic Woman, at ask@techtalk.digital. TECH TALK 183: 31 JAN 2020

PORT COURT REPORT

Do me a favour. Go to your desk or office or wherever it is that you keep your electronic goods. Open the bottom drawer. Move aside all the old mobiles and electronic devices you have tucked away in there. Now find the cables. How many do you have there and, more importantly, how many different connection types do you have?

If the answer is less than four types, I will be surprised.

Micro-USB; Lightning; USB-C; Mini-USB; USB-3; 30-pin and so many more including many proprietary connectors.

Each manufacturer claimed that their connector was superior and it was needed to achieve the full performance of the device but the cynics in our society often saw it as either a revenue raiser or a way to lock you into one brand.

One of the reasons we end up with so many connectors is that every new device comes with a charger. Imagine the negative publicity that a company would receive if it started selling a device where the cable was NOT included.

It is annoying on Christmas Day when the new toys you purchase for your kids don't include batteries but imagine if the new phone you bought had no charger? Well that is exactly what the Members of the European Parliament have just voted in favour of. With a vote of 582 to 40 the resolution for a single universal charging solution was a comprehensive victory for those in favour of this solution to reduce e-waste.

This has been a campaign waged for the last decade with an estimation that obsolete cables generate more than 51,000 tonnes of waste per year.

When the campaign started, there were more than thirty different charging solutions. In effect, the resolution decouples the sales of devices and chargers. What a concept.

On the plus side, it should marginally reduce the cost of our new purchases (granted – the chargers don't cost the manufacturers a lot) and it should reduce the clutter in our drawers. I can also see innovative charging solutions with devices that charge multiple devices with a common connector. It also helps when you arrive at work or visit a friend and need a charger – you know there is going to be a common charger available.

On the negative side it has been argued (mainly by Apple, who have the most to lose) that it will stifle innovation. Apple has typically had a proprietary charger (think of 30-pin and now Lightning) so it is another way for consumers to be locked into their ecosystem. You may accept Apple's argument if their solution was light years ahead but when you consider the data transfer rate of a Lightning port is capable of about 480Mbps compared to that of USB-C which is 10Gbps or Thunderbolt which is 40Gbps then the argument falls a little flat.

Next steps? Assuming all goes well, it will eventually be mandatory for all devices sold in Europe to have the agreed upon charging port. Logically, once that occurs, I would assume that manufacturers will adopt that standard for all devices sold across the world. From a manufacturing perspective, it would not make a lot of sense to make a device for Europe and a device for the rest of the world.

With this victory, I would now strongly encourage the decoupling of remote controls from televisions. I am not sure what I have more of – old chargers or old remotes. Imagine a universal remote control that all televisions used and a purchase of a TV did not come with another remote. Tell me if you would vote for that at ask@techtalk.digital. TECH TALK 184: 7 FEB 2020

Next Text Specs

What do you do when you manufacture a television with specifications so high that there isn't content at the same level? The obvious answer is to manufacture a phone that can create content at the same level!

Samsung has a range of 8K QLED televisions from 55" up to the $1K per inch 98" version. One of the problems I have discussed previously with 8K televisions is the lack of content. Most of the loungeroom viewing across the world at the moment is in HD with a smattering of 4K.

Noting this 8K dilemma for Samsung, at the latest Samsung product launch, all three new Samsung phones will be able to shoot video in 8K and even be able to stream 8K phone footage direct to an 8K TV. Clever cross-marketing between their phone and entertainment divisions!

I am always excited at the launch of a new phone. Each new product pushes the technology envelope just that little further and exciting new features quickly become commonplace across many mobiles.

What did Samsung offer to entice us to upgrade? Apart from many minor improvements, the battleground was clearly defined as camera and connectivity. There were three new models announced by Samsung in the Galaxy S20 range. The S20; S20+ and S20 Ultra with screen sizes of 6.2"; 6.7" and 6.9" respectively. By way of comparison, the three iPhone 11 models are 6.1" for the 11 and the Pro comes in 5.8" and 6.5".

Sizes seem to be on a never-ending upward trend which is great for useability but makes pocketing the device awkward. If only we could just fold up the device and put it in our pocket!

Samsung introduced 5G connectivity last year which was a little like their 8K television problem. A 5G phone when very few networks across the world have moved to 5G is a lot like a 4G phone.

Samsung are of the belief that 5G will start to progress at a much quicker rate this year so two models come in a 4G and 5G version while the Ultra is only available in 5G. Apple's latest products offer approximately zero 5G options but I can't see how they won't change this when they release their next flagships in September this year.

Then there is the camera – or more accurately cameras. I have mentioned the 8K video footage but the raw image size of the still photos is impressive. I don't want to get carried away by raw image size alone but...WOW! In amongst the four cameras on the S20 Ultra sits a 108MP camera. Even the front-facing 'selfie' camera is 40MP. Despite many AI improvements, digital zoom is never as good as optical zoom but Samsung are talking up their 100x zoom on this phone. A demonstration at the launch of a picture of Alcatraz prison from a cliff two kilometres away certainly looked impressive and surely they wouldn't use any trickery at a launch, right? On raw specs Apple seems to be dragging the chain. The best they offer is 12MP on their cameras and 10x digital zoom. There is more to a camera than just the number of pixels and the iPhone cameras are still impressive but...108MP? WOW!

Samsung have firmly placed the ball in Apple's court and it is now up to Apple to return the ball - and I haven't even mentioned Samsung's two folding phones. While I am excited about this competition from a technology perspective, the ultimate winners are consumers. Bring it on! Tell me the next killer feature you would like to see in a phone at ask@techtalk.digital. TECH TALK 185: 14 FEB 2020

Neat Seat Feat

Ahhhh – the thrill of attending a live sports match. Fighting my way through city traffic and lining up early to secure the best seats in the stadium. Lathering on sun screen throughout the day. Lining up to get a beer that is almost as good as one sitting at home in the fridge. Lining up in the food queues in the hope that my favourite version of Gözleme hasn't sold out. Arguing with the slightly inebriated person who seems to have stolen my seat after lining up for my food. Missing some crucial action on the field while still focused on said inebriated person. Crossing my legs tighter so I don't miss any more of the action...It doesn't get much better than that!

Except...

A few of my friends choose to sit at home to watch their sport. They commute 30 seconds to their lounge room. They watch their 85" TV with every angle imaginable – including cameras on the players. They have ready access to food and toilets all while relaxing at a controlled comfortable temperature.

It is no surprise that consumers are choosing to sit at home and watch major events rather than attend in person.

But technology is providing a potential reversal – in the form of smart stadiums. Yes, it is trendy to place the word 'smart' in front of anything and hope that people are attracted but there truly are some innovations that will change the way we view sports.

At the heart of this revolution is connectivity. Both in Wi-Fi 6 and 5G.

Neat Seat Feat

There is no doubt that the atmosphere at a live event is better than your lounge room at home – no matter how many noisy friends you have with you. But the viewing experience is usually somewhat diminished. The first solution? Broadcast throughout the stadium the same, or better, footage than viewers see at home. No subscription required to a pay TV operator. If you are in the stadium you can use your smartphone to view the action – and most importantly look at the replays. Add in some ability to change the view to your preferred angle – including live on the players – and it sounds attractive.

This development will also see Augmented Reality which will allow you to point your phone at a certain area of the field and see a crystal-clear view of the action at that point.

Forget about lining up at the food vendors. Use the stadium app to order and pay for your food – which will be delivered to your seat! Entry will also be made easier. We have the ability now to use our phones to scan our tickets as we enter a stadium. Next step? Facial recognition. When you purchase your tickets, upload your face and simply stroll through the facial recognition line.

No need to worry about a flat battery – all seats come with a USB port and wireless charging to keep you juiced up throughout the match and some stadiums even have heated seats with a 4D option to match the action on the field. A bone-jarring tackle? You will literally feel it as well.

Smart stadiums with some of these features are already being rolled out across the world as we speak – look up the Tottenham Hotspur Stadium; AT&T Stadium and SoFi Stadium as examples - but as the connectivity improves expect to see more innovation in this space.

The last resolution that I am after – how can I avoid a trip to the toilet? Solve that one and the solution is complete. Tell me if you prefer a stadium or your home at ask@techtalk.digital. Tech Talk 186: 21 Feb 2020

EXAGGERATE EQUIVOCATE AUTOMATE

Many people would be familiar with the name Alan Turing from the 2014 historical drama, *The Imitation Game*, which was the highest grossing film of that year. Turing is widely considered to be the father of artificial intelligence and I remember studying one of his concepts at University. It was known as the *Turing test*.

This is a test of a machine's ability to exhibit intelligent behaviour indistinguishable from that of a human. The test came from his paper that posed the simple question: "Can machines think?" Turing went on to explain that 'thinking' is difficult to define so posed the solution to the problem with the idea of a game called the 'imitation game.' If a person has a text only conversation between either a computer or a person and can't reliably tell the machine from the human, then the machine has passed the test.

Turing's paper was written in 1950 when the majority of communications occurred verbally – mostly in person. Telephones were available but, in the US, there was only one telephone for every five people. Some at the time dismissed the Turing test as not a relevant test of 'thinking' because it removed voice from the conversation and relied on a text only conversation. When would humans have a conversation just with text?

Fifty-six years later along came Twitter. The perfect platform for Turing's test.

Along with texting and e-mail and other social media platforms, think of how many people you have had conversations with that have been purely via text.

Luckily we know that we are having those conversations with real people and we could definitely tell the difference if it was just a computer at the other end.

Or could we?

Researchers have found that bots are used extensively on Twitter to amplify a message. 38 per cent of tweets related to 'fake science' were from bots. 28 per cent of tweets related to Exxon were not from humans and 25 per cent of climate change denial tweets were bot generated. An increase in social media traffic is a way that many people, and even media organisations, determine that there is public interest in a particular topic. It is a principle known as 'social proof.' It is why we are attracted to a restaurant that seems busy despite the fact that we would probably receive our meal quicker in an empty restaurant. If lots of other people are eating there, it must be good! Right?

If an organisation created a method to influence public opinion just by tweeting more, surely that could have a major impact on the bottom line. But it would be unethical so that would never happen. Think again. It was discovered last year that the mining company, Glencore, spent up to £7 million per year with the C|T Group over several years with teams in Sydney and London to spread anti-renewables messages and counter anti-coal activists.

The latest evidence shows that Glencore was charged way too much. Other companies are using bots – at a much cheaper rate – to deliver the same social media influencing.

There is no doubt that Turing was a visionary and way ahead of his time but I am not sure that even Turing could have imagined that his theoretical test would actually be used every day by billions across the world.

Next time you read comments or interact with someone on social media that you have never met, ask yourself if you think it is actually a human or a bot - and then how you really know.

Send any further thoughts to ask@techtalk.digital – humans only please. TECH TALK 187: 28 FEB 2020

Dire Inquire Misfire

I recently had an experience I would not wish upon my worst enemy (I don't actually have enemies – but the expression sounds very dramatic!) My wife and I were sitting at home relaxing on a Saturday afternoon. Our four children were at various locations away from home including our youngest daughter on a student exchange in Japan.

Our blissful peace was interrupted by our gate bell. We checked the video intercom and two uniformed officers were at our gate.

The officers told us they were here to help as they had received a triple zero emergency call. We assured the officers we hadn't made the call and they then quoted a mobile number.

Our hearts sank. The mobile was the number of one of our daughters. Not the kind of news you want. You imagine the worst possible scenario with your daughter trying to ring for help. A very quick phone call to our daughter was met with great relief when we found out she was enjoying a day at the races and had managed to accidentally hold down a key combination that made an emergency call. The officers came to our house – not the race track which was hundreds of kilometres away - because that was the registered address of our daughter's mobile.

Most people would assume that the 'authorities' have the ability to track your phone if they needed to do so. If you watch a few movies you might also be under the impression that the 'authorities' can also use a satellite to zoom in on a number plate on a moving car and hit a magic 'enhance' button.

For the people that made the 8.9 million calls to triple zero last year, they would be well aware of the fact that the operators have no idea where your phone is located. The operator will ask for an exact location including the nearest cross street. Pretty tough if you are on a country road!

That is set to change as Australia will shortly see the rollout of automatic mobile location (AML) technology that will allow location accuracy within three metres.

The technology involves collaboration with the manufacturers and telecommunication providers and will be managed by the Department of Communications and the Arts. This is still a work in process but I would expect to see the connections completed by the end of this year. The end result will allow a triple zero operator to receive a call from a person using a mobile phone and almost immediately know their location. This allows the operator to find out more about the issue and send first responders to the exact location. As an added bonus, it will mean that the thirty per cent of non-emergency and prank calls to triple zero will reduce. If a caller knows that their location can be tracked to such accurate levels, they may think twice before calling triple zero to order a pizza or ask the time.

While this all sounds fantastic and will save lives, there is always the potential for the number of organisations accessing the information to grow. I am quite comfortable with triple zero knowing my location when I make a call, but I would prefer Harness Racing NSW and Bankstown City Council didn't know my exact location. These were just two organisations who previously requested access to telecommunications data without a warrant. Those requests were specifically related to metadata but it will require vigilance to ensure AML data doesn't suffer from scope creep.

Tell me how you think we could improve the current triple zero service at ask@techtalk.digital. TECH TALK 188: 6 MAR 2020

Skip Trip Microchip

If you were one of the tens of millions of people around the world who watched a day in the life of Jack Bauer unfold in the television series, '24', then undoubtedly your highlight would have been similar to mine.

Day Six, which aired through 2007, was the first time that most people saw the concept of sophisticated videoconferencing. Cisco paid a ridiculous amount of money for their technology to be used as part of the plot when Noah Daniels and his staff communicated with Yuri Suvarov and the Russian staff using the newly released Cisco TelePresence suite. Product placement in movies and TV shows is so prevalent now that audiences are very sceptical of brand names in shows. 'Modern Family' went as far as basing an entire show around the Dunphy family scrambling to buy their gadget-loving Dad, Phil, a new iPad for his birthday. The first iPad wasn't released until three days after the show aired!

Thirteen years ago, though, Cisco was able to place their products in a hit TV series and the motivation was never questioned. What this did, along with many other similar product placements, was usher in a world where we believed that the general concept of videoconferencing was about to take over the world. Conference centres would close down; airlines would go broke; muffin-producing cafés would lose their best revenue source. The roads leading into major cities would be pleasant and free-flowing as employees sat in their homes and performed their work.

What a beautiful image.

It is a shame it was only ever portrayed on marketing brochures and not in the real world.

Until…

Along came a pandemic that has directly impacted 125,865 lives so far with the tragic loss of 4,615 of those people.

Suddenly Mobile World Congress was cancelled. Then Adobe Summit. Next Facebook and Google cancelled conferences. IBM, Microsoft and TED all followed suit. It was just too risky to have a large gathering of people when Coronavirus (COVID-19) posed too great a risk of infection. Large organisations are asking their staff to telecommute.

If any good can come from the heartbreaking loss of so many lives, it may be that we finally learn to embrace the concept of videoconferencing. That can bring significant benefits to society in terms of reduced congestion on our roads, reduced need for office space and reduced emissions with lower levels of travel required.

There is still a way to go though. Barriers in the past have all been loosely related to quality. If you watched '24' and saw the quality of their conference calls, you would rush out and set up the equipment immediately – but, and I apologise for any shock I may cause here – but TV shows and movies aren't always portraying things with 100 per cent accuracy.

The opportunity is now there for the current organisations that provide videoconferencing services – such as Lifesize; Skype; Zoom; Cisco WebEx; GoToMeeting – and new competitors to come in to the market to capitalise on the current set of circumstances. Telecommunication providers around the world can respond to the challenge with greater bandwidth availability and governments could even design initiatives and incentives to encourage activity in this space. It may be as simple as holding an ad-hoc FaceTime meeting with 32 people on an iPhone all the way through to dedicated rooms for your videoconferencing applications.

There is an argument that some of the best work done at a conference is around the bar at midnight so tell me if you would be happy to forgo your next physical conference for a virtual conference at ask@techtalk.digital. Tech Talk 189: 13 Mar 2020

SWELL CELL EXPEL

It was almost two years ago when I almost gave my daughter a heart attack. Our family was on an Easter holiday and one daughter was at University so she flew direct to our destination. We were about to prepare a meal in our apartment when I picked up my daughter's phone – and realised that her short time at University had resulted in a phone covered with residual alcohol stickiness.

So, as any good father with a knowledge of technology would do, I turned on the tap and placed her phone under the slow stream of water. I don't think Shelly-Ann Fraser-Pryce could move as fast as I saw my daughter move. She imagined all of her photos and contacts and apps and, well, her life, disappearing down the drain and bounded across the apartment.

> The phone had an IP67 rating and a slow stream of water was not going to cause any damage.

Sony was the first to introduce a modern smartphone with a water resistance rating. In March 2013 the Sony Xperia Z was released which had an IP57 rating. Samsung followed with the Galaxy S7 in March 2016 with an IP68 rating and Apple came to the party with the iPhone 7 in September 2016 with a slightly watered down (sorry – I couldn't resist a second pun) IP67 rating.

The "Ingress Protection" ratings are defined in EN 60529. The first number relates to resistance to dust. A 6 means that the product is dust tight. The second number relates to water resistance. A 7 means that the product can be immersed in up to 1 metre of fresh water for up to 30 minutes. A rating of 8 goes a little further with the manufacturer required to specify the exact depth. For example, the iPhone 11 Pro has an IP68 rating and Apple specifies the phone can be submerged in up to 4 metres of water for up to 30 minutes.

As a handy side benefit to a little bit of water being splashed around, with COVID-19 permeating every aspect of our lives, a phone that is water resistant is a phone that is easier to clean.

The phone is almost a perfect breeding ground for the spread of any disease. Apart from most millennials, the rest of us talk on our phone. Our phone is close to our mouth so we are breathing our germs all over it. Then we swipe our fingers across a germ-ridden screen. Our hands are now ready to transfer these germs to the world. I would add cleaning your phone to the list of precautionary measures to take.

Firstly I would check the IP rating on your phone and, if it is IP67 or IP68 rated, I would start with a brief rinse under slow running water. Keep in mind that there is risk involved with this as manufacturers still won't cover water ingress under warranty.

Secondly I would use any common substance that would be used to clean a pair of glasses. Spray the substance onto a lint-free cloth and then wipe the phone with that. Apple advise that a 70 per cent isopropyl alcohol wipe, for example, is perfectly fine to be wiped over the non-porous surfaces of the phone. Don't be too aggressive though – many screens have an oil-repellent coating to minimise finger smudges.

Most importantly, don't spray anything directly onto the phone. Always spray onto a cloth and then use that to wipe over the phone. These are strange times indeed. Tell me if you feel comfortable putting your IP rated phone under a tap at ask@techtalk.digital. Tech Talk 190: 20 Mar 2020

Satellite Height Finite

Many people talk about the pace of change 'these days' but I think it would be fair to say that COVID-19 has delivered a world where the pace of change is not measured in years or months but in hours and minutes. Just when we think we have somewhat of a handle on how the world will look going forward, the ground shifts from underneath us.

From a technology perspective, I spend the majority of my days helping clients improve their remote connectivity to satisfy their requirements in working or studying remotely. If there is any positive from COVID-19 it may be that, in the future, we have an entirely different approach to our idea of where the workplace is. Less commuting would mean benefits for humans and the environment.

As a result of people staying at home, our Internet traffic across the entire world has increased by 8 per cent with some parts of the world seeing an increase of 50 per cent. This is placing incredible stress on the entire Internet and, although it is coping at this stage, it is certainly under stress. Major streaming providers have tried to help with the problem by reducing the quality of their streaming services. It may seem a backward step but the lower quality streaming has a dramatic overall impact on the strain placed on the Internet.

Apart from people having time to watch streaming TV while they sit at home and worry about whether to pay their rent or buy food, the main use of the Internet is to allow people to try and continue on with a normal life while physically located at home. That can involve working with video or tele conferencing and, in general, accessing their normal work tools while at a remote location.

With the advent of the NBN, you would think that Australia is perfectly positioned for remote connections. Yes and no. Let me start first of all with a major bugbear I have with our current NBN solution. I remember way back in 2009 when I was sitting in the office of then Communications Minister, Stephen Conroy. Satellite was part of the overall new drive forward with modern communications but I argued satellite was just not good enough. At a height of 35,786km above the earth and the maximum speed of electromagnetic waves, as first proposed by James Maxwell in 1865, the inherent latency of 477 milliseconds renders any experience with satellite involving two-way communications near useless.

Fine for watching movies but remote working becomes problematic.

For people currently using satellite I typically recommend they use external antennas or signal boosters or stand on one leg and hang their tongue out or use any method possible to gain access to mobile broadband and then use a dedicated mobile broadband device to gain access to the Internet. Unfortunately, the plans are not as attractive as fixed NBN services but, depending on reception, the speeds are usually quite reasonable.

The connection is the crucial component but, once you have that finalised, it is then a matter of deciding what you need to do while you are remote and what tools you need. Will generic tools – word processing, spreadsheets for example - cover all your needs or do you have proprietary tools that you need to use. Are these located in your workplace or are they available in the cloud. One huge advantage of businesses that have moved many services to the cloud is that accessing those services is easier from a variety of locations.

Tell me what technology you need to work remotely at ask@techtalk.digital. TECH TALK 191: 27 MAR 2020

Vacation Simulation Creation

Do you remember the old days? The days when we could jump in a car and drive to a local tourist destination and gather with other people whilst viewing the attraction? Or cast your mind back further – when we could fly across the seas and visit a foreign land.

Once at a faraway location, we had the pleasure of patiently standing in a line in the sun/rain/snow/heat/cold (insert applicable for your favourite destination) and paying good money to gather amongst the crowds and see a glimpse of the attraction. When I saw the Mona Lisa (which I can only *assume* was the original) I was surrounded by 200 of my closest acquaintances – and that was in the quiet season!

With the new world we now live in, all of those pleasures have been removed. No more visiting foreign destinations. No more being stuck on a plane beside a crying baby. No more jet lag. For the time being at least.

None of this removes the fact that there are some fascinating destinations to visit. For anyone that has run into a friend just after they have returned from a holiday, by the time they have finished showing you their photos and videos from their trip it feels like you have been on the trip with them – without the cost.

As with most problems I see in life, I assume technology is the answer. For lack of travel caused by COVID-19, I see virtual holidays and tours as the answer.

My favourite destinations to visit virtually are a combination of real and imaginary.

The first virtual tour I would recommend is the TARDIS. The most famous Police Box in the world only appears with an external area of 3.9 square metres. Inside is a different story. Make sure you have some time on your hands because on the inside the TARDIS contains an infinite number of rooms, corridors and storage spaces.

For a nerd like myself, there is an attraction in taking a virtual tour of CERN, the European Organisation for Nuclear Research. Probably best known for the location of the Large Hadron Collider, the facilities host a huge range of scientific testing equipment.

Along more traditional lines, the Roman Colosseum is a popular destination – both in the real and the virtual world. Virtual tours have one major advantage in allowing you to go to some sections of a destination that are out of bounds for you when you visit in the physical sense. Bonus content!

The International Space Station (ISS) is a location that most of us are unlikely to visit with a total of just 228 people having ever been to the ISS. Although astronauts don't need to wear a spacesuit inside the ISS, to really feel like an astronaut I imagine wearing several layers of clothing to restrict my movement, wear a facemask and climb under my bed before taking this virtual tour.

Another location that is out of reach for most is Mount Everest. Only 4,042 people have reached the summit – but with a virtual tour we can all experience the scenery minus the cold and lack of oxygen.

Other notable mentions in the list of tours to consider include The Louvre; the Kennedy Space Centre; The Sistine Chapel; Antarctica; the Great Wall of China; Pyramids; the Grand Canyon; Google Data centres and for a final fictional one, Diagon Alley.

They may not be quite the same as the real thing but at the moment, what choice do we have? Send me the list of locations you would like to take a virtual tour to at ask@techtalk.digital. Tech Talk 192: 3 Apr 2020

Input Output Throughput

NBN recently released information to demonstrate just how much additional data we have been using since the start of COVID-19 restrictions. In less than a month our usage in the evening has increased by 25 per cent; the early evening has witnessed an increase of 30 per cent and 21 per cent is how much more data we are seeing during business hours. This doesn't include increased data on the mobile phone network. To give you an idea of the sheer volume of data, the evening now sees a peak download rate of 13.8 terabits per second (Tbps) on the NBN with 9pm the peak time for usage.

It would be easy to dismiss our slow Internet as another COVID-19 related issue and just accept it.

I think we can do better than that.

The first step is to understand what is happening when you use the Internet and where potential bottlenecks may be. I like to explain the Internet like a garden hose. When you turn on your sprinkler the amount of water that comes out is determined firstly by your actual sprinkler. But there is also your garden hose and your tap and the pipes in your home. Then there is the usage in the home. Are the kids in the shower? Is the dishwasher going? Then there is the size of your water meter and it then flows into your street where once again the pipes installed by your water authority control the volume. Your neighbours then play a part with their usage and so it goes all the way back to the water treatment plant. Each step can impact that sprinkler flow but you have less control the further away the issue is from your home.

The Internet can be thought of in a similar fashion.

Firstly download two apps. Speedtest to test speeds to the outside world and Wi-Fi SweetSpots to test your internal Wi-Fi.

Firstly, look at what devices might be generating Internet traffic. Do you have devices that are automatically backing up data to the cloud? Do you have computers infected with viruses or malware? The best method is to turn off all devices and do some Speedtest testing with just one device at a time turned on. If you notice a big drop in speed with one particular device on, have a closer look at that device.

Most devices in the house are connected via Wi-Fi and this is an area that is well within your control to potentially improve. Most Wi-Fi devices have the ability to work on two frequencies – 5GHz (not to be confused with 5G) and 2.4GHz. 5GHz should be faster but 2.4GHz will give you better range. The way to test is to use Wi-Fi SweetSpots as you move to different parts of your house. Try testing between both frequencies. Also experiment with different locations within your house of your actual Wireless Access Point or Router. Placing the device higher or in a location away from metallic objects (fridges, filing cabinets) will often give you better range.

Lastly there is your Internet Service Provider (ISP) to consider. They will often have different plans or speeds or even consider changing to a different ISP who may give you better performance. Even though different providers may quote the same speed they may not have the ability to deliver on that speed when traffic increases. As with life, it is usually a case of you get what you pay for. The dearer plans and providers usually provide better performance.

Tell me how much you have noticed your speed change at ask@techtalk.digital.
Tech Talk 193: 10 Apr 2020

Reengineer Gear Fear

The earliest archaeological evidence of any type of wheel was discovered in Mesopotamia dated around 5,500 years old. Those early wheels were used for pottery creation so it was left to the Ancient Greeks to develop the wheel for transportation in some basic axle carts. The wheel has developed somewhat since those early wheelbarrows and, despite the fact that the US Patents Office alone has issued more than ten million patents, the wheel is generally regarded as the greatest invention ever.

Hence the oft-used expression that we don't need to 'reinvent the wheel' when it comes to creating something that has already been created.

I am concerned that this is the exact path the Australian Government is currently heading down with the announcement that a Coronavirus Tracing App is currently under development. This is on the back of a similar app currently being used in Singapore (called TraceTogether) and other apps being developed in the US and England.

As the promotional video for TraceTogether states, only two steps are required. Download the app and turn on Bluetooth. In a culture that is largely compliant, less than twenty per cent of the adults in Singapore have downloaded the app.

The Prime Minister has set a target of forty percent installation rate to make the app effective. "Tell him he's dreamin'" because if Singapore can't achieve twenty then we have no hope of forty. Various governments are missing the obvious opportunity that has presented itself by two of the largest companies in the world. In a rare move, Apple and Alphabet (the parent company of Google) have collaborated to present a solution to try and help the world beat COVID-19. If there was one single action that sent a message of the severity of this virus, collaboration between these two competitors is it.

These two companies have developed a solution that allows the operating systems of Apple and Android phones to exchange data when phones are nearby using low-range and low-power Bluetooth signals. The location is not recorded and neither is any personal information. The record of these nearby interactions is kept for only fourteen days.

If a person is diagnosed with COVID-19 they will need to opt-in to the system to notify any phones they have been near in the past fortnight. Most importantly, the operating system itself will alert people that they have been exposed to someone who has subsequently been diagnosed with COVID-19. This is crucial. It doesn't rely on each user downloading an app. It also doesn't tell people when or where the exposure was therefore protecting the privacy of the individuals.

There are still some issues. It still relies on people having a smartphone – and with only a 77 per cent penetration rate in the over-60 age bracket, there is the potential to miss many of our most vulnerable people. It also relies on users upgrading to the latest version of their phone's operating system but the success rate with this is much higher than the success rate of installing a new government sponsored app. Privacy issues are also a concern. Do people have more trust in their government or behemoth technology companies? Unfortunately, there isn't really time to debate this issue. If we can save lives, we may need to place some faith in these organisations and scrutinise their actions during and after the process.

Technology can indeed help with the defeat of COVID-19 but governments need to work smarter and collaborate with organisations across the world.

Tell me who you would trust more with your data – large technology organisations or the government – at ask@techtalk.digital. Tech Talk 194: 17 Apr 2020

Wrecked Inspect Correct

The first car I owned was a V8 253 Holden HZ Ute - although I am ignoring my brother's VC Valiant which he accused me of borrowing so often that I thought I owned it! I had gained most of my mechanical skills working on motorbikes as I was growing up but, after having owned the HZ for a number of years and visiting auto-wreckers around the nation, I was well and truly acquainted with the inner workings of the 253 (and the Valiant).

In our modern world, it is unlikely that I would have been allowed to touch either vehicle as I certainly wasn't an authorised repairer – but that may be about to change with the help of some Aussie farmers.

> There is a 'Right to Repair' movement that has been gathering momentum around the world.

The general logic from a manufacturer of high-tech equipment is that only the manufacturer (or authorised repairers) has the technical skill and know-how to repair their equipment. Consequently, organisations around the world work hard to gain 'authorised repairer' status for a variety of manufacturers. It isn't quite a monopoly, but if you perform repairs yourself or use someone other than an authorised repairer, your warranty will be null and void. It tends to fly in the face of open competition.

The 'Right to Repair' concept started in the automotive industry in the US when Massachusetts passed the Motor Vehicle Owners' Right to Repair Act in 2012. It took six years for the automobile trade organisations across the US to finally sign a memorandum that said they would provide the necessary documents and information to allow anyone to repair their vehicles.

You may remember a similar process last year that focused on mobile phones and electronics in general. The trigger was battery replacement in modern smartphones. Many phones were being replaced and ending up in landfill for no other reason than the fact that the battery life had diminished to the point where the phone was not useful. Replacing a battery would not only void the warranty but, in some cases, render the phone useless.

The change in laws for electronics would not only allow products to be repaired but would require manufacturers to make products that could easily be repaired.

Enter the good old Aussie farmer. The simple tractor is not so simple any more. They can cost close to a million dollars and with GPS and self-driving technology along with soil monitoring, they are a sophisticated technological masterpiece. That sometimes breaks down. When they do break down, the farmers are held to ransom – both in terms of time and money – by the authorised repairers of the manufacturer.

The ACCC has been brought in to help with the battle and an inquiry has started. Farmers will have the chance to put their case forward to be able to choose who repairs their equipment – including the option of themselves. Despite manufacturers saying that they are prepared to play the game, independent repairers still complain that they don't have access to parts or information to be able to carry out repairs.

It can even drill down to the ownership of the data. With information recorded by farmers trying to maximise the efficiency of their operation, a change of manufacturer may render historical data useless – effectively locking a farmer into one brand forever.

Keep an eye out for our farmers making changes to the way technology is developed and repaired across the world – with a little help from the ACCC.

Tell me if you think the HZ or the Valiant was a better car at ask@techtalk.digital.
Tech Talk 195: 24 Apr 2020

Sick? Pick Tick!

I walked in to a local sporting goods store many years ago to buy a sports watch. The typical sort of thing for the day. A chunky watch with a separate external armband to track GPS location and a chest band to monitor heart rate. I remember asking the salesperson if there were plans to incorporate, at the very least, the heart rate function into a watch on your wrist and possibly even GPS functionality.

The salesperson was quite confident that a watch would *NEVER* have the capabilities to monitor heart rate and a GPS was way too large to be incorporated into a watch.

I learnt a long time ago that using the word never in terms of technology is dangerous – and so it would be the case in this scenario.

If I could have jumped in my time machine (borrowed from Leonard and Sheldon) and brought back a watch from the future to show this particular salesperson, I can't imagine the reaction when the features of a modern smartwatch were discussed.

The latest feature that caught my attention was the ability for a watch to measure blood pressure. Most people would be familiar with the process of measuring blood pressure. A sphygmomanometer is used with a cuff and stethoscope to measure the systolic and diastolic pressure.

How can that measurement process be delivered with just a watch? There is a small trick – an electronic cuff is used once every 4 weeks to calibrate the watch but, apart from those calibrations, a watch on a wrist is all that is needed.

I have been watching (sorry – couldn't help it) the development of wrist technology with interest and, even involved in technology, I find it incredible what the humble wristwatch is capable of. My Mum was absolutely gobsmacked the day I showed myself and my daughter making calls between our two watches.

As kids we used to joke about having a watch like Dick Tracy but here we were actually using this technology that was first dreamt of in cartoons.

The combination of cellular capabilities and GPS and motion sensors delivers an incredible safety feature. Fall detection. Some modern watches will detect a fall and, if the watch detects no further motion for a minute, the watch will automatically call emergency services. If the watch detects movement after a large impact, it will still ask you to call emergency services but await your response.

There is a focus on health with many of the watches so there is no surprise that watches will detect heart rate; blood oxygen levels, sleep patterns and some can even allow you to take an ECG. From a practical perspective, having the ability to use mapping capabilities on the phone itself makes it seem like a GPS in a car is so old-fashioned! The list of features continues with notifications from social media and other apps popping up on your watch in addition to other alerts that may be anywhere from an irregular heart rate to notice that someone is at your front door.

Listening to music with Bluetooth headphones has never been easier and I still see retail assistants give me a smile of childhood awe when I use my watch to pay for items. With all of these features, most smartwatches are capable of being underwater to a depth of 50 metres – which finally allows the metaphor to be tested when someone is said to be able to talk underwater with a mouth full of marbles.

Tell me the next killer feature you would like to see in your next smartwatch at ask@techtalk.digital. TECH TALK 196: 01 MAY 2020

Invitation Admiration Damnation

As humans we have a basic desire to be loved – and an innate curiosity. If you want to see fast actions from a person, mix these two ingredients and stand back!

"This is a tech column," I hear you thinking. "Why is he talking about psychology?"

This last week we saw the twentieth anniversary of one of the most infamous computer viruses the world has ever seen – and it relied on these two aspects of human behaviour to be so devastating.

Looking back now the 'Love Bug' seemed so incredibly simple and basic. Filipino Onel de Guzman, aged twenty-four at the time, sent out an e-mail with an attachment. The subject line of the e-mail was "ILOVEYOU" and it had an attachment which was called "LOVE-LETTER-FOR-YOU.TXT.vbs"

What brilliant social engineering!

Someone is in love with me? What have they put in the letter? In some circumstances you knew the person who sent it to you or it may have been a name you didn't immediately recognise. Is it the person in accounts that says hello to me in a very friendly manner? Someone I see playing sport on weekends? What about that person behind the bar at the pub? We always have a few laughs.

Most people couldn't wait to open the attachment to find out more details about the person who was smitten with them – but to their disappointment nothing was there. Nothing they noticed anyway – except their computer was now infected with a virus and, more importantly, everyone in their address book was sent a copy of the virus. *You* were now professing *your* love for someone else, and on it went. To approximately forty-five million computers around the world.

Guzman has said that, as a young student, all he was after was free Internet access. He wanted to steal some passwords so he could use the credentials that someone else was paying for. Certainly not pure motives but not in his wildest dreams did he imagine that he would shut down the UK Parliament and impact the Pentagon.

The Melissa virus was released the previous year and it infected a million machines but that was nothing compared to the Love Bug.

Computer viruses have a long history in the world of IT. The first documented virus was the Creeper program which was released way back in 1971. It was not malicious and was really just a test of a self-replicating program.

As computers developed, some programmers liked to experiment with ideas. In 1981, Elk Cloner used floppy disks that the Apple operating system was stored on to infect Apple II computers and in 1986 the first IBM PC compatible virus was launched. Over the ensuing years various viruses were launched with some level of notoriety – such as Michelangelo; Ping Pong; Father Christmas; CIH and more. In relative terms, these were not widespread. What a virus needed was a better transmission tool.

And that tool was the Internet.

Since the Love Bug we have seen many major outbreaks such as Code Red and MyDoom and only three years ago the WannaCry ransomware attack. The world is certainly more aware of the potential damage that can be done and antivirus software is more sophisticated – but the programmers are also more advanced in their attacks.

And what happened to Guzman, the writer of arguably the most infectious virus the world has seen? It was a case of legislation being out of date. There was no law against malware in the Philippines at the time so he was never prosecuted.

Tell me if you feel safe against computer viruses at ask@techtalk.digital. TECH TALK 197: 8 MAY 2020

SHARE CARE ANYWHERE

When I was young, I would sometimes sit with my hip and groovy (much) older teenage sisters and listen to their music – usually on records but sometimes just listening to the radio. I remember one night when I heard Kiki Dee's unique soul vocals pump out "I've Got the Music in Me" and at the end of the song, the radio announcer declared that she must have swallowed a radio!

It was a poor joke but we found it amusing and rang the radio station to tell the DJ of our amusement. His response was interesting. He told us that he sits in the studio with a record player and microphone and talks to himself. If he didn't have the microphone in front of him then he may be committed to an institution for the same actions. He appreciated the fact that someone out there was actually listening.

He had a good point. You can look at radio ratings and estimated listener numbers and perform surveys but, as a radio host, when you are talking, right at this precise moment in time, is anyone listening?

When someone sets out for a career in radio, they are presumably aware of this issue but educators are different. When your career path involves standing in a school classroom or a University lecture hall, you fully expect to be able to see the whites of their eyes. Well that was yesterday. Today is a new world. Students all over the world are currently sitting at home. For the educator, accustomed to seeing people in front of them, the best they may have is a thumbnail on screen of each student. Pretty tough to gauge the engagement level of the audience.

Until now.

A data analytics team is currently working on developing a program to give real-time feedback to a teacher on the engagement level of students sitting in front of their computer. By tracking eye movements; facial expressions; head actions along with mouse and keyboard activity, researchers are determining the amount of engagement each student has with the current delivery of information. By taking the combined data from all of the students in one class the teacher sees a line graph of their current engagement level. The line starts to go up – fantastic! The students are really listening to what I am saying. Line starts to drop? I guess they aren't interested in my Boomer anecdote. The greatest advantage is that, just like with face-to-face teaching, the teachers can subtly and continually alter their delivery to make sure they are reaching their students.

With theories abounding on what our PCV (post coronavirus) world will look like, it is generally accepted that online learning will be a part of that world. Current research indicates that 46 per cent of young students decrease learning outcomes in an online world.

This technology is only in its infancy and just 500 hours of data has been logged so far but COVID-19 gives researchers a rare opportunity to gather vast amounts of data to make giant leaps in this technology.

One of the positives that we may see PCV is for educational opportunities to be available to students regardless of their postcode.

The next scary step with the research? Imagine a badge that you could wear to tell the person you are speaking with your engagement level. Over 25 years of marriage I think I have managed to feign interest in my wife's shoe shopping expeditions but if technology started to be introduced, I may be in trouble!

Tell me your view of education PCV at ask@techtalk.digital. TECH TALK 198: 15 MAY 2020

Beside Graveside Denied

It may seem like quite an easy feat in our current environment but way back in 2006 I had an interesting request from a friend. His elderly father had passed away and their family was spread far and wide around the globe. Unfortunately, it simply wasn't possible for one particular sibling to make it home from overseas. The friend asked me to setup a live stream of the service.

It wasn't called a live stream back then but the concept was the same. I somehow felt slightly awkward sitting at the front of the church configuring my technology to achieve the desired outcome with the ancient technology solutions available at the time – but the family appreciated the inclusion of their sibling from afar.

Fast forward to 2020 and the technology has improved dramatically. For a start, the term live streaming is understood by...well...everybody. The ability to deliver high upload speeds with the consistency required to send video has also improved. The camera quality has improved as well and there are now a plethora of devices with high quality cameras and connectivity.

With the technological advances there was one thing that didn't change. The awkward feeling at a funeral if someone setup some technology that live streamed a service.

Until COVID-19 came along.

When our Government limited the number of mourners to just ten, funeral directors had a different playing field. As much as I don't want to think too much about my funeral for many years, I certainly hope that I have more than ten people at my funeral.

Many funeral directors across the nation are now offering the ability to live stream a funeral. The initial feedback from funeral homes is that families are initially not sure about the concept. Some families are not sure it is showing the right amount of respect for the deceased – but they are taking a leap of faith with such tight restrictions in place.

With some funeral homes seeing over 250 people attend a funeral in a virtual sense, friends and family are obviously feeling comfortable with the concept. Funeral homes have been aware of the ability to provide this service but, until COVID-19, were conscious that it might not be received that well by families. The feeling so far is that, even after restrictions are lifted, this will now be a standard service that many funeral directors will offer to their clients.

How times change.

It made me start thinking about some other potential positives from this current pandemic. We have been discussing and talking about and postulating on remote work for many years. The technology was definitely there – but the will wasn't. Suddenly we have had people embracing the concept of sitting in their business shirt and board shorts and enjoying the serenity of their homes in a relaxed manner after not having endured the commute to work. Maybe this commute from the bedroom to the office will take off? Not only does it have the potential to increase productivity but the lowered pollution has been noticeable.

Educational opportunities have also been conspicuously increased. With school and university students adapting to remote learning as quickly as educational institutions have had to change their practices, it suddenly opens up education for people regardless of location. Sure, there are still some limitations to work through. Certain courses that involve a practical component – such as medicine and engineering – present harder challenges but the speed with which we have adapted to our new environment gives me an incredible feeling of hope.

Tell me how many people you would like to attend your funeral at ask@techtalk.digital. Tech Talk 199: 22 May 2020

FLEW THROUGH TWO (HUNDRED)

16 April 2006. Chittagong. The first day of the last Test in the Australia versus Bangladesh series. Jason 'Dizzy' Gillespie shared the new ball with Brett Lee. After Gillespie had finished his first five overs, he had the figures of three for eleven and wasn't used again in the innings. The Tigers were all out for 197 late in the day and Dizzy had the best bowling figures and best economy rate for the Aussies.

When Australia went out to bat, Gillespie thought he was finished for the day but, as he tells the story, "Ricky Ponting, the world's best batsman, tapped me on the shoulder and, after having just routed the opposition with my bowling, asked me to do *his* job as well. He just felt that the Bangladesh attack posed too many threats to risk him at number three." He was given the role as 'nightwatchman' and when Matty Hayden holed out, Dizzy went out to face the last seven overs and protect his captain.

Not only did he do that but by Day 4 of the Test, on his 31st birthday, Gillespie reminded the Bangladesh players of the original Dizzy Gillespie's 1988 album when he batted *Endlessly* to reach the most unlikely of double centuries. The only nightwatchman to do so ever. The only player in the history of Test cricket with an average below twenty with a single score above 200.

He was named Player of the Match and the Series – and was never selected to play for Australia again.

In May 2016, then editor of an Australian Fairfax publication, the late Brian O'Flaherty, published my first Tech Talk column. I had been writing IT columns for IT audiences since 2004 but this was my first serious foray into the general populace. Today sees the publication of my 200th column so, just like Dizzy, a double century that I wasn't necessarily planning upon when Brian asked me to submit my first column. This column now has a home in 140 mastheads.

When I look back over those two hundred columns, the variety of topics explored has been vast and the changes in technology? Immense.

My columns have explored mobile radiation and SAR measurement through to current 5G conspiracy theories. I have discussed methods of international roaming and Internet access everywhere from homes to planes. Cars are a favourite – electric; hydrogen; autonomous; driver assisted and more. My nostalgic bent gives me the opportunity to reminisce about video cassettes and vinyl records and my old BlackBerry through to games such as Wolfenstein and Crash Bandicoot or the tech used by 007. Democracy solutions aren't missed with my ideas on electronic voting and how social media sites need to address misinformation and even how content moderators are now playing the role of editors.

Rumours on new technology launches are revealed and then, after launches, the latest products are dissected to give you the latest information to help make decisions on your technology. Announcements at international conferences are broken down into what may be relevant to the real world. Technology gift ideas are explored – they may be Christmas gift ideas or what I am buying my wife for her birthday.

Advice is given to help prevent ransomware or phishing or scams or just how to setup home automation.

If it is technology I can see society still talking about and using in five years then it has a place in a Tech Talk column. Hopefully, unlike Dizzy, I will continue on to my triple century and well beyond.

Tell me a topic you would like to see discussed in a future column at ask@techtalk.digital. TECH TALK 200: 29 MAY 2020

Substation Conservation Domination

I have often mused about the lack of intellectual discourse and logical debate around election time. Three-word catchphrases and headlines seem to dominate the landscape which doesn't leave a lot of room to demonstrate the development of an idea or a plan.

Maybe that suits some political parties who would be exposed if more than a three-word catchphrase was required – but I digress.

Electricity has been a recent election battleground where philosophically opposing views have met practical considerations of turning on the lights.

While the pollies are busy arguing, industry itself will work out the direction. I delivered one of my futurist presentations back in 2017 when I was describing our world in the year 2025. One of my data points was that I predicted that wind and solar would make up over twenty per cent of power production by the year 2025. It was a brave prediction at the time.

The Australian Government has just released information on how our energy was produced last year. Our total energy generation was 265.1TWh – an overall increase of 0.8 per cent. Given our population growth was 1.5 per cent last year, this is an indication of power saving initiatives bearing some fruit.

Coal is still king in Australia with 56.4 per cent of our power produced by burning coal but this has seen a drop over the last year of 4.5 per cent. This is a continuing trend with the reliance on coal reducing by 7.8 per cent over a four-year timeframe. No real surprises with the second largest source of power – gas. It makes up 20.5 per cent of our production which is an increase of 5.8 per cent year on year and a 6.8 per cent four-year increase.

Renewables are where most of the action is at the moment. Wind blows hardest in the renewable space with 7.4 per cent of our power produced with wind turbines. This is a healthy 19 per cent increase over the last year and 64.9 per cent over four years. Solar currently sits at 6.8 per cent of our overall power production but it has increased by 46.2 per cent over the last year and 189.2 per cent over four years. All that rooftop solar is having an impact.

Since 1974 when the Snowy Mountains Scheme in NSW was officially completed, hydro has been an important part of the power mix in this country. Last year generation surprisingly fell by 17.5 per cent although, with a four-year rise of 1.6 per cent, I suspect last year was an anomaly and we will see more power flowing this year.

If we break it down to non-renewable power versus renewable power, we start to see a definite trend. Reliance on non-renewable power dropped by 1.7 per cent last year and renewables increased by 11.5 per cent – admittedly from a much lower base. Take a four-year view and you see a 4.4 per cent drop in non-renewables and a massive 54.4 per cent increase in renewable power.

What is the current mix between the two production methods? Drum roll please...Non-renewable power now sits at 79.1 per cent of our production and therefore renewables sit at 20.9 per cent. And what of my 2017 prediction on wind and solar? We are at 14.1 per cent so well on the way to my predicted twenty per cent and, based on current trends, the 50 per cent renewable power by 2030 is no longer a prediction – it is stating the obvious!

Tell me what year you think we will hit fifty per cent of renewable power production at ask@techtalk.digital. Tech Talk 201: 15 Jun 2020

KEEN GREEN QUEEN

As I walked through my local supermarket this week, I performed one simple action that reminded me of the incredibly opulent society we live in. I checked the time on my watch and noticed that it was a little chilly outside. I removed my phone from my pocket and opened my car's app. I turned on the artificial climate inside my car and cranked it up to a comfortable 23 and turned the seat heater on. When I arrived at my car it would then already be at my preferred temperature and hence save me a couple of minutes of slight discomfort!

Well that was certainly an essential use of the hundreds of billions invested over thirty-eight years in connecting devices around the world as part of the Internet of Things (IoT)!

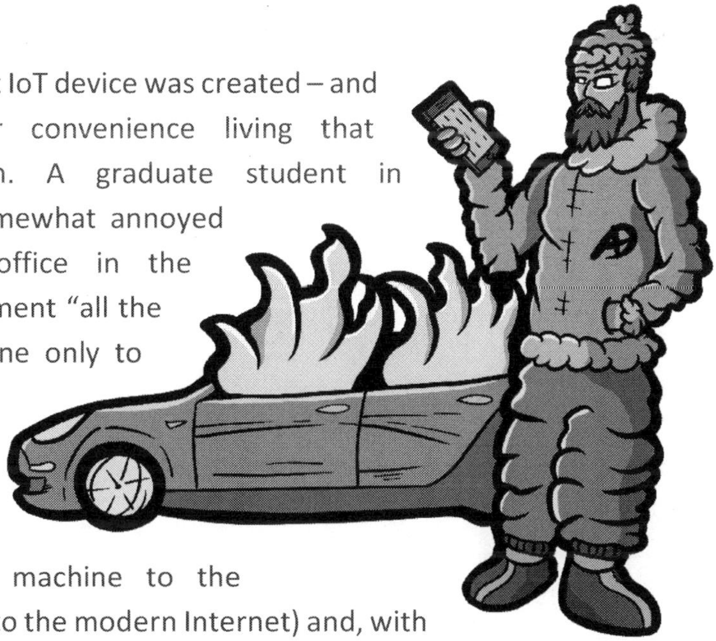

It was in 1982 that the first IoT device was created – and it was through similar convenience living that prompted its invention. A graduate student in Pennsylvania became somewhat annoyed at walking from his office in the computer science department "all the way" to the Coke machine only to find it empty or, worse still, the bottles inside were tragically warm. The solution? He connected the vending machine to the ARPANET (the pre-cursor to the modern Internet) and, with some clever programming, he could check from afar the availability and temperature status of the drinks.

Even though the term IoT was not coined until 1999, this machine was the first device connected.

The IoT market is currently exploding with over 30 billion devices connected to the Internet with estimations of up to 1 trillion devices connected by 2025.

Cars are a simple example but what else is happening in this space?

Smart homes and smart appliances will be the largest sector. In an example of what is available right now, I already have the ability to control and monitor many aspects of my home. Air-conditioning; door locks; security system; doorbell; cameras; gates; garage doors and lights. My kids have grown up in a world where you use your smart phone to turn lights on or off or, more commonly now, they simply ask Alexa to do that. Appliances with in-built IoT will be the next wave in the home.

Asset tracking will be another huge use of IoT. Companies with fleets of vehicles will be able to track the location and speed of any vehicle in their fleet. I had a very interesting situation many years ago when I helped a client keep track of the whereabouts of a wayward employee using photos from a digital camera. Today it would be as simple as tracking your fleet with IoT sensors. Using that IoT data in the public domain allows real-time traffic congestion in major cities to be incredibly accurate. Asset tracking can vary from a vehicle up to a road train carrying freight down to the individual freight parcels or, in our case, down to my daughter's favourite teddy bear. Seeing the precise location of your pizza or taxi is one thing but seeing the accurate location of your parcel takes it to an entirely new level.

IoT devices will continue to proliferate in manufacturing and robotics; healthcare facilities; wearable devices; smart meters and the farming sector but what I am really waiting for is an IoT connected ball! With an object only 42.67mm in diameter and able to experience a G force of over 20,000, it is a complicated problem – but still one easier to solve than keeping my drives on the fairway!

Tell me what device you would like to see IoT connected at ask@techtalk.digital.

Tech Talk 202: 12 Jun 2020

Cite Bright Sunlight

Way back in 2009 I was writing columns for a computer magazine and I was discussing some concepts to reduce our carbon emissions. One of my speculative ideas was to produce all of Australia's power via solar panels. My rudimentary calculations at the time showed that we would only need about one thousand square kilometres of solar panels to power this nation. That could be in one location (think of our Simpson Desert) or distributed (think of rooftop solar) but that was the total area required.

We have moved forward eleven years since I wrote that and the landscape has certainly changed. In 2009 we used 244TWh of electricity. This year we will use approximately 265TWh. An increase of only 8.6 per cent despite a population increase of 19.2 per cent over that same time.

So based on consumption, the amount of power needed would not need to increase dramatically since my 2009 calculation.

The area that has changed dramatically though is the technology in the solar panels themselves.

> At the equator at midday the surface of the earth receives approximately one kilowatt per square metre of energy from the sun.

Converting this to electricity is the trick. In 1839 when the photovoltaic effect was discovered, production of electricity was not even at 1 per cent efficiency. By 1954 the efficiency had only reached 4 per cent. Today we see panel efficiencies as high as 22 per cent with the theoretical limit for silicon PV efficiency calculated to be 29 per cent. When you consider that this nation averages 16.1 million terawatt hours of annual radiation even at 20% efficiency that is about 12,000 times more power than we use. We don't want to cover every square kilometre of our nation with solar panels though but we are being quite creative with what we are doing.

Australia leads the way in rooftop solar panels with the highest per capita rate of installations in the world – double the rate of the next closest nation.

One issue that has been highlighted is the consumption of raw materials in the production of solar panels. Silver is an integral component of a modern solar panel with a massive 15 per cent of the world's industrial consumption of silver used in the production of solar panels. Silver is a finite resource but some clever Aussie ingenuity is working on a solution. Copper is 1,000 times more abundant and 100 times cheaper than Silver but has not been able to be used in solar panels – until now. One Australian company has created a process that will see Silver replaced with Copper while maintaining efficiency rates. The most exciting part is not just the innovation and advancement of an already well-developed technology but the salivating prospect of an entire manufacturing process being created in this country. With so many solar panels being used across this nation, it is possible that the new post-COVID-19 economy features a reduction in globalisation and perhaps even an increase in manufacturing in Australia. When you combine well-known Aussie ingenuity with local manufacturing you even have the possibility of an export market.

A recent Ernst and Young report laid out the details of a renewables led recovery from COVID-19 and that also featured increased manufacturing so with continuing improvements in technology and a desire across the world to produce more power using renewable sources, don't be surprised to see more solar panels in our future – with the exciting bonus of a sticker that says 'Made in Australia' on them.

Tell me if you have rooftop solar panels on your house at ask@techtalk.digital. TECH TALK 203: 19 JUN 2020

Wannabe Key Guarantee

I was a car thief once. OK, so not actually a car thief, but I remember the day when I certainly looked like a car thief. I didn't have a dark hoody and gloves that seem important for car thieves in the movies but I did have the one piece of equipment that was important. Blue packing tape...and I was also caught in the act of breaking in to a locked car.

The major difference between myself and a car thief, and what I had to explain to my new friends in blue, was that the car I was trying to unlock was my Mum's car and ten minutes earlier she had given me the car keys. The same car keys that I could now see sitting on the seat where I had accidentally left them.

Back in the eighties it was easy to lock your keys in the car.

> Roadside assistance stats from that era showed that being locked out was more common than a flat battery; running out of fuel or a flat tyre.

As I sat through the Apple Worldwide Developer Conference keynote address at 3am earlier this week one of the many announcements made me reminisce about my brush with the law.

The Apple Digital Car Key was demonstrated during the virtual conference. Hold your phone near your car door to gain entry and then place your phone on a pad inside the car and push the start button. Accessing your car has been a technological battleground over the years. In 1980 the Ford Thunderbird introduced SecuriCode – a 5-digit keypad mounted above the door handle. The

Renault Fuego introduced the first handheld transmitter on a car in 1982 and by 1998 the Mercedes Benz W220 started the smart key revolution.

The concept of using a smartphone to access your car is not entirely new. I own a car that is permanently connected to the Internet and I can use my smartphone to not only open and start my car but access it from anywhere in the world.

What made the Apple announcement significant is that the technology used is NFC (Near Field Communication). If you use a card or smartphone to tap for payments you are already using NFC. NFC is designed to work under a distance of 4cm and provide a fast and secure connection between two devices.

Remember the days when you measured the importance of a person by the conglomeration of keys hanging from their belt? I envisage a day when keys will be no more. Not everyone is comfortable connecting their car and house locks to the Internet but when the unlocking device has to be within 4cm of the lock, the comfort level increases dramatically. I already use NFC for locks in my business so this isn't some futuristic idea that is being dreamt about in laboratories – it is already happening.

Jump forward a few years. Your smartphone will be your universal key. It will be the way you unlock your house, your business, your car. You will use it to pay for items. Log on to your computer. Cash withdrawals. Vending machines. Flight check-ins. Medicare.

Convenient? Yes. Scary? A bit.

My takeaway from the Apple announcement? Maybe not today, but certainly as we go forward, make sure your phone is always charged! Probably a good idea to have a reliable backup as well. On a bigger picture, we will have to have a lot of trust in technology companies to get the security right.

Tell me if you would feel comfortable replacing your physical keys with your smartphone at ask@techtalk.digital. Tech Talk 204: 26 Jun 2020

UTILITY SURVIVABILITY FACILITY

It was the night before my twenty-first birthday party and I was in high spirits. For some inexplicable reason, it was decided we should test the idiom of "zìxīang máodùn" or more simply put, what happens when an unstoppable force meets an immovable object?

In this case the unstoppable force was my V8 253 Holden HZ Ute and the immovable object was a 6.5m Ironbark pole with a 300mm radius that was buried 2m into the ground.

As any good scientist would tell you, the answer to a theoretical question is experiment.

The result of the testing was that the immovable object remained largely intact and I am only here to tell you the tale because of the work of Swedish engineer Nils Bohlin in 1958 and a 1970 announcement by the Chief Secretary of the Victorian State Government, Sir Arthur Rylah.

Technology had saved my life and the lives of two friends in the car. The modern seat belt was invented by Bohlin whilst working for Volvo and the car manufacturer then generously gave it to the world, saying it was too significant as a life-saving tool. The Victorian Government was the first in the world to make the wearing of a seat belt compulsory.

In 1929 and 2018 the road toll in Australia was almost identical – 1,145 and 1,135 - but the highest fatality year, both in outright numbers and per capita, was 1970. 3,798 people lost their lives at an average of 30.4 per 100,000 of our population.

When you look at a graph of our road toll over the last century, there are obvious changes when technology has advanced or legislation has been changed.

The introduction of radial tyres in 1948 saw the road toll show major improvements for a decade. The seat belt in 1970 has been the greatest single factor in the history of the car to reduce the fatality rate. The introduction of random breath testing in 1982 delivered immediate benefits. Australian Design Rule 69 was introduced in 1995 and made air bags compulsory but they had already started to be introduced by various manufacturers over the previous fifteen years.

We may shake our heads with the benefit of hindsight, but with each technological breakthrough, car manufacturers typically ignored the option at first, then offered it as an expensive option or in their high-end vehicles before either governments or market forces deemed the product compulsory.

With some incredible safety options now available for a modern car, one research study has actually placed a figure on how many lives are being lost with the technology delay. The study concluded that if it was compulsory for all new cars to have all the latest safety technology installed, fifty-eight per cent of current fatalities would be avoided.

The types of technology we are talking about here are extensive. Electronic stability control; improved braking systems including anti-lock braking systems and electronic brake-force distribution and one of my favourites, automatic emergency braking; lane departure warning systems (or lane keep); blind spot detection; facial alertness recognition; traction control; active cruise control; intelligent speed assist; pedestrian detection systems; vehicle to vehicle communications; drink driving prevention technologies; adaptive cruise control and many more.

Of course, these systems cost money. If all of these systems were deployed in every new car, the estimation is that it may increase the price of a new car by up to $15,000 which, particularly at the entry level, is a significant sum of money. But what price a human life?

Tell me the one technology item you believe should be compulsory in new cars at ask@techtalk.digital. TECH TALK 205: 3 JUL 2020

Keen Teen Screen

"According to most studies, people's number-one fear is public speaking. Number two is death...Now this means to the average person, if you have to go to a funeral, you're better off in the casket than doing the eulogy." Jerry Seinfeld, in his introduction to Episode 63 of his popular sitcom series, noted there are seemingly endless surveys and studies of everything from, well, your number-one fear to whether a good playlist will help you work out harder (apparently it does).

As both a parent and a technologist, I was interested in a recent study that sought to discover if parents are more comfortable talking to their children about the birds and the bees or the bits and the bytes.

> As awkward as it may seem, fifty-six per cent of parents would prefer to talk to their children about the mechanics of reproduction rather than reveal their lack of knowledge when it comes to technology.

Furthermore, sixty per cent of parents said they use their children for advice on technology. There was no figure for how many parents use their children for advice on sex...

One real challenge for any modern parent is just keeping up with the number of acronyms. In any industry, initialisms and acronyms are subtly used to let people know who is in 'the club' with a blank look quickly appearing on the faces of those who are outsiders. Even when we think we are keeping up to date, the rules quickly change with no memo from head office to help us out. I distinctly remember one awkward situation when I was coaching my daughter's netball team and one of the teenage players sent me a message which she signed off LOL. In my day LOL meant Lots Of Love. I was a little embarrassed and confused by this casual use of a term of endearment. Upon asking my daughter for advice she politely advised me, after she recovered from laughing, that LOL meant Laugh Out Loud. But no one told me! She did LOL!

If you flip the tables around, kids feel about the same way as their parents. Forty per cent of the kids who participated in the survey believe they are more educated on technology than their parents and when they do need advice, fifty-three per cent of them turn to their peers for help.

One of the greatest strengths of technology is also its greatest weakness. Technology, by its very nature, moves quickly and changes constantly. Unlike the proverbial bike that you never forget how to ride, if you look the other way with technology, it may be unrecognisable when you look back again.

Staying up to date with technology requires an attitude of learning and curiosity. Whether you are seven of seventy, if you find the world interesting and want to know more, you will be able to stay in touch with technology. If you prefer to live your life in set and forget mode, be prepared for your kids to quickly bypass your knowledge.

With a quarter of kids seeking information from YouTube and sixty per cent of children prepared to seek advice from their elders - if only they knew more - the obvious answer is for Mum and Dad to create their own YouTube channel giving advice to kids! Either their kids would love to tune in and seek advice – or they would be so embarrassed that they would agree to continue to talk to Mum and Dad if they would just take down that silly YouTube channel!

Tell me if you still give your kids technology advice at ask@techtalk.digital. Tech Talk 206: 10 Jul 2020

Seal Wheel Deal

Way back in those crazed eighties when most kids were watching a Steven Spielberg movie on their Betamax or learning to moonwalk like Michael Jackson, I was doing my best impression of emulating the stars of BMX Bandits - although I never could find a girl that looked like Nicole Kidman collecting trolleys at the local supermarket.

We would visit Tom Dawson at his cycle shop and marvel at the latest Mongoose and Quicksilver BMX bikes – in sizes small and large - and Tom would look at our height and help us adjust the seat height and head stem for maximum efficiency.

Several decades later and not a lot has changed. Buying a modern bike involves measuring up some vital statistics, choosing a frame size and having an expert at your local cycle shop set the seat height and head stem at the correct settings to suit your body.

But that is all about to change.

Superstrata Ion is the first bicycle manufacturer in the world to create each bike based specifically on a rider's height; weight; arm length and riding style. After various measurements are taken the details are fed into their CAD software and a frame that is specific to you is created on screen – and then 3D printed. Made from carbon fibre, of course, the frame is lightweight and strong and, most importantly, customised specifically for you. Add the required components to the customised frame and you ride off into the sunset using your available power in the most efficient manner.

Not since Suntour invented the slant-parallelogram rear derailleur in 1964 has the

bicycle industry undergone such a revolution as that afforded by carbon-fibre. Forget Steel; Reynolds 531 tubing; Duraluminium or even Titanium. Carbon-fibre offers such an incredible strength to weight ratio – approximately ten times that of common steel alloys - that designers often throw out the traditional double-triangle design and start again. The 3D printed carbon-fibre bicycle is the latest iteration of development aided by this wonder material.

Gaining an extra few per cent efficiency from your bicycle is one thing but the 3D printing revolution is opening up an incredible world of development.

The basic concept seems simple. Everything that we know of on this planet is made up of 118 elements. Combine those elements in the correct fashion and you can 'print' anything. And manufacturers are doing just that.

The International Space Station has a 3D printer on board to print spare parts that may be needed. Why carry huge inventory in a limited space when you can just print what is required. Tools and musical instruments are being 3D printed. Food is being created and biology is now being developed with prosthetics, human body parts and even human organs. The list seems endless: skin; buildings; clothes; cars; shoes...even 3D printers are being made by using 3D printers!

At this stage, manufacturing high volumes of a product is still more cost efficient using traditional methods rather than 3D printing but the introduction of 3D printing has two significant advantages. Firstly, the production of customised or short-run products is faster and cheaper. Think of the Superstrata Ion as an example. A bike customised just for me using traditional methods would be slow and expensive (two words that may be used to describe me!) Secondly, manufacturers are now using 3D printing for rapid prototyping. A design can go from concept to prototype in a matter of hours not months. This helps bring new designs and concepts to mass market much quicker and at a lower price.

Tell me what you dream of 3D printing at ask@techtalk.digital. TECH TALK 207: 17 JUL 2020

Background Sound Drowned

A decade ago I was privileged to have a tour through a new $18 million tiered theatre that was nearing completion. What most impressed me was the acoustic performance of the theatre. Sitting in the audience in any seat a person on stage could be heard clearly with no amplification necessary but then when the half million-dollar sound system was fired up, the acoustics were even more impressive.

At least I thought so.

One of the other people in the tour group was a self-confessed audiophile and he told everyone who cared to listen about the distortion and flutter of the sound system and said that it needed hours of work to fine tune the expensive equipment.

If I didn't know it before then, it was obvious to me that I was no audio expert. Even after hearing about all the supposed problems in the sound quality, it still sounded pretty impressive to me.

It was similar when vinyl fanatics dismissed the Compact Disc (CD) albums when they were introduced as being too 'cold' compared to the 'warmth' of a vinyl record.

My experience with vinyl was that records were often

scratched or had dust on them and they popped and crackled like Coco Pops.

One thing that even my limited high-fidelity hearing can attest to though is that the cassette tape definitely is poorer sound quality. The cassette was introduced by Philips in 1963 but became wildly popular through the seventies and eighties with the cassette used in everything from a 'boom box' to car stereos and the biggest consumer hit of the decade, the Walkman which was released in 1979 and inspired the 1981 Cliff Richards hit, 'Wired for Sound'.

The largest issue with the cassette tape is simply lack of data availability. The tape itself is only 3.81mm wide and that contains four magnetic tracks – a stereo track in each direction. The tape then only moves past the read heads at 47.625mm per second. Add in some stretching and dust and chewed tapes and the CD couldn't come fast enough. 1993 was the year that CD sales overtook cassette sales.

So why am I even discussing cassettes twenty-seven years after they were deposed as the king? Well the king is making a comeback. In the UK in the first half of this year, 65,000 cassette albums were sold. My immediate thought was that these were old nostalgia albums but Lady Gaga sold 12,000 of her latest album on cassette. Overall sales have doubled compared to the same period in 2019. There are some things in life I just don't get and this is one of them. Vinyl and CDs and streaming are all better quality than cassettes. Finding a song on a cassette is a clumsy process of fast forwarding and hoping – even vinyl is easier. Streamed music won't skip when used in a portable device. CDs don't degrade in quality. I am really struggling to see why anyone would want to buy a cassette.

It just goes to show. For all the research and reading and predicting that I do, when it comes to technology, predicting the future is incredibly difficult because every now and again, you see something like this increase in cassette sales and it just doesn't make sense.

Oh well, if you can't beat 'em, you may as well join 'em. I am sure I have an old Walkman and an eighties mix tape somewhere in my bottom drawer. I am off to do some roller-skating!

Please tell me any reason why you would buy music on a cassette at ask@techtalk.digital. Tech Talk 208: 24 Jul 2020

Known Phone Zone

Now normally I don't like to blow my own trumpet but...Toot Toot! This week saw the thirtieth anniversary of the mobile phone network being switched on in my home city. My business had the privilege of enabling the very first connection in that city. I had started my first electronics business the previous year but it was on a trip to the Adelaide Grand Prix in November 1989 that school friend, Daryl Coxon, told me all about this new concept of mobiles. He said there was probably a market in it for me.

Now, with almost 80,000 mobile phone sales under my belt, I think Daryl was right!

> It was in 1973 that Dr Martin Cooper, working for Motorola, famously called his chief competitor, Dr Joel Engel at AT&T. "Joel, this is Marty. I'm calling you from a cell phone."

In August 1981, Telecom Australia launched the Public Automatic Mobile Phone System (PAMTS) with three towers in Melbourne. The mobile numbers started with 007 which seemed appropriate for the peak of 14,000 subscribers as the entire concept seemed like something from the movies. This system was shut down in 1993.

The Advanced Mobile Phone System (AMPS) was switched on in February 1987 with an extravagant event at the Opera House with skydivers and Government Ministers. Most of the early models were installed in vehicles with the option of a portable 'bag phone' or a too-big-to-fit-in-your-hand handheld version. This 1G network was shut down in September 2000. The first phone I sold was an NEC P3 handheld with a phone number of 018 63 62 62. Optus and Vodafone entered the Australian market in 1992 and the three carriers had a combined peak of 2.6 million analogue subscribers.

GSM (Global System for Mobile Communications) was the standard developed by the European Telecommunications Standards Institute and was chosen by Australia when it was switched on in April 1993 and enjoyed a twenty-five-year life and a peak of 15.9 million users. This was 2G and was our first 'digital' system. While the service offered additional features, such as text messaging, the range wasn't as extensive as AMPS. CDMA was introduced in September 1999 mainly to cater for regional users. CDMA offered range similar to analogue with services similar to digital and reached a peak of 1.8 million subscribers before being closed down in July 2012.

The number of mobiles in Australia reached a significant milestone in February 2001 - 10.7 million active services - surpassing the number of landlines.

As data services became more important, the search for speed led to the 3G network. It was turned on in Australia in April 2003 and was a preview of how our phone usage would start to change. Better data speeds allowed different functionality which eventually led to the smartphone. The 3G network is slated for shut down in Australia in 2024.

After the BlackBerry and then the revolution of the iPhone announcement in 2007, the pressure of data traffic led to the launch of the 4G network in September 2011. This is still the network that most Australians are using today...until they buy their new phone to take advantage of the increased speed and reduced latency offered by the 5G network which launched in May 2019. We currently have over 30 million combined active mobile services in Australia – more than the number of people.

It has been an exciting journey to witness the introduction of a range of services at the coal face over those last thirty years.

Tell me when you started using your first mobile – and the model - at ask@techtalk.digital. TECH TALK 209: 31 JUL 2020

Hardware Warfare Fanfare

I attended a local Chamber of Commerce function in April 2004 and the guest speaker was Tom Potter. No, not some distant relative of Harry. Tom founded the Eagle Boys Pizza chain in 1987. During his presentation, Tom mentioned that they had expanded the chain through New Zealand and had a monopoly on certain aspects of the pizza business in that country.

As a throwaway line, Tom noted that many people think competition is good in business but he was of the opinion that owning a monopoly business is so much better!

Jump forward twelve years and I was attending another local Chamber of Commerce function and this time, Don Meij, the CEO of Domino's, was speaking. Don believed that competition drove any business to greater heights and spoke at length about their innovative plans to stay ahead of the competition with various technology strategies including drone delivery; app ordering; voice assistants and artificial intelligence.

Within a few months of that function, it was no surprise to me that administrators were brought in to the Eagle Boys head office while Domino's proved that drone delivery was no 'pie in the sky' concept with the world's first drone delivery.

One aspect of our current technology banquet is that we have wonderful competition amongst three very strong players. I am sure Samsung; Apple and Huawei would all prefer to have a complete monopoly on the smartphone market but, as a consumer, I love the competition.

I was up in the early hours this week to watch Samsung Unpacked – their biannual event to announce their latest products. Every company is trying to emulate the presentation style of the late Steve Jobs standing in front of screaming Apple devotees but this event was particularly awkward. With no audience save the 'Samsung Members' who were attending virtually, it had a very stiff feel about it.

Ignore the presentation style though - what about the products? Samsung announced five new products. Samsung typically jostle for the number one position across the world in overall smartphone sales – but Apple still outsells them in the premium space. The Samsung Note 20, announced this week, is another step forward in that space. The Note 20 has the best camera ever in a smartphone and the S Pen is more responsive than ever before. The 240Hz screen is perfect for gaming – which is convenient because Samsung partnered with Microsoft Xbox Game Streaming. Samsung has had a bumpy start to the beginning of the foldable phone era but this week the Galaxy Z Fold 2 will hope to win more fans with a 6.2-inch outer screen folding out – hopefully with no issues – to 7.6-inches. My ultimate phone would be the physical size of a watch but, just like the TARDIS, be dimensionally transcendental so it is shown to be larger than it is. Given that there is a position open at Samsung at the moment for a transdimensional engineer, the flip phone is the closest we have. While Samsung sits around the top of the smartphone market, Apple owns the Tablet and Watch space with 38 per cent and 56 per cent market share respectively, Samsung released four new versions of these devices. With the Tablets boasting 5G and functionality matched to beauty, they will definitely convert some current iPad users. The two watches now match Apple for features, add on 1mm each in size and throw in blood pressure monitoring and an ECG for good measure.

I love being a consumer in these technology wars!

Tell me the killer feature you want to see in your next phone at ask@techtalk.digital. Tech Talk 210: 7 Aug 2020

Great Slate Fate

I started selling computers in 1989. Desktop computers. It took me three years to sell my first notebook. Despite the fact that desktops only hold a market share position of 22.6 per cent now, thirty years ago it was rare to see someone use a notebook. There were two main reasons – high price and poor performance.

The first notebook I did sell had a monochrome screen and was using the brand new 486SX processor along with 2MB of RAM and an 80MB hard drive.

And I wanted one – but they were way too expensive for me to even contemplate. My dream at the time was to make enough money to afford a notebook!

The brand of that first notebook? Toshiba of course. Toshiba was the first company to sell a laptop PC. In 1985 they launched the T1100. It sold for $10,000 in current money and was a little ahead of its time. It ran MS-DOS and had the new 3.5-inch floppy disk drive. Pundits at the time thought it was too early to launch a portable product but the head of Toshiba's PC business believed in portability and convinced his superiors that he could achieve 10,000 sales a year. At their peak, Toshiba sold 17.7 million notebooks annually!

Throughout their history of notebooks, Toshiba often led the way. I did manage to finally buy my own Toshiba notebook and Toshiba was my brand of choice for about fifteen years. I was very proud of my Satellite R20 which I purchased in 2006. It was a tablet notebook when most people used the term tablet to refer to medicinal pills. With a traditional notebook format as well as the ability to spin the screen on top of itself and use a stylus plus a docking station to use it as a replacement desktop, it truly was a three in one device.

But alas, in a lesson that will be shared in business schools for several decades, even the mighty can fall. After thirty-five years in the notebook business, many as the leading brand in the world, Toshiba notebooks are no more. This week witnessed the sale of the final twenty per cent of the Toshiba notebook division to Sharp. In 2018 Sharp purchased eighty per cent of the business for a miserable $36 million. When you consider WhatsApp was bought by Facebook for $19 billion, it does seem like loose change.

Despite a reputation for innovation, in the end it was the changing market that saw the demise of Toshiba notebooks. Market share for notebooks is currently almost twice that of desktop PCs with 42.4 per cent but Toshiba's sales fell from 17.7 million a decade ago to less than a million in 2018. The surge in popularity of the iPad and other tablets was part of the problem. Tablets took out a 35 per cent market share last year – but the reality was that more players crowded the market and Toshiba lost their reputation as market leaders. Quite the opposite in fact. After fifteen years of relying on Toshiba for my personal notebook, there came a day a few years ago when I simply couldn't ignore the fact that buying a Toshiba was tantamount to buying yesterday's technology. And so I didn't.

To quote supercoach Jack Gibson, "If you are standing still, you're going backwards fast" and never is that more obvious than in the technology space.

The successful technology businesses of tomorrow are the innovators of today – the ones pushing the boundaries to see what is possible.

Tell me if you ever owned a Toshiba notebook at ask@techtalk.digital. Tech Talk 211: 14 Aug 2020

EXPECT CONFECT EFFECT

Pick your extravagance of choice. Chocolate, coffee, fruit, whatever. If I offered you a free single serving of your indulgence immediately OR a full day with free access to your preference if you went without for a week, would you wait for the greater reward?

This is the basic premise of the Stanford Marshmallow Experiment. Young children sat down with one marshmallow in front of them. If they could resist eating the single marshmallow for fifteen minutes, they would receive two. In follow-up studies, those that resisted had better life outcomes.

We all struggle to plan for tomorrow because we see our future selves as strangers. The problems of tomorrow can be dealt with by someone else – even if that someone else is us in the future. It is why we have compulsory superannuation and why COVID-19 has resulted in more government action than Climate Change.

Most people acknowledge that electric cars are our future. In my role as a technologist and futurist, I was recently asked when I thought electric cars (EVs) would achieve price equivalence with petrol cars (ICEVs). I told the audience we are already there! The outbursts were instant with people quoting ticket prices of EVs versus equivalent ICEVs. This was a perfect example of the present ignoring the future.

The largest single cost of owning a motor car is not the purchase price. It is the depreciation cost. This is followed by registration and other fixed costs and then running costs such as fuel and maintenance. The actual purchase price is not the only factor you should be looking at. If people used the same logic when considering their home, we would not have the 67 per cent level of home ownership we have in this nation. Rent is cheaper than the huge price you pay for a home so logically everyone would keep renting!

Assume I am about to purchase a new small vehicle for my daily commute to work. This segment makes up 15.4 per cent of our one million annual sales. I am trying to decide between a $30,000 ICEV and a $50,000 EV. The EV *seems* significantly dearer but assume the following: Five-year ownership; 70 per cent depreciation of ICEV versus EV 50 per cent; rego and CTP the same but insurance dearer on the EV; annual service costs lower on the EV; fuel price average of $1.55/litre over the next five years and average electricity cost of 10c/kWh (combination of free charging stations and off-peak).

The fixed annual costs are $5,900 for an ICEV versus $7,100 for an EV. Variable costs, such as fuel and maintenance, delivers a per kilometre cost of 12.4 cents per km for an ICEV versus 3.28 cents for an EV. Multiply those costs by the 13,400km ABS average for all Australian vehicles and the total is $1,662 versus $440. Bring it all together and you have a total annual cost of ownership of $7,562 for an ICEV versus $7,540 for the EV. The gap widens for every kilometre you travel over the national average. If you look at the overall cost of owning a vehicle, admittedly with many assumptions that could vary over the next five years, the price of an EV is on the money.

I like to walk the walk and my household is now proudly ICEV free. Having owned a Tesla for almost three years I have now added a 2020 Nissan LEAF to our garage to replace a hybrid vehicle.

I believe the electric future is here now!

Tell me if your next purchase will be an EV at ask@techtalk.digital. TECH TALK 212: 21 AUG 2020

GREAT CELEBRATE DATE

I was looking through some photos recently and stumbled across an old photo of me from 1995. It was a newspaper clipping that had me with my head down on my keyboard sound asleep. No, it wasn't a photo at the end of a night of hard-core gaming. It was at the opening of my business at midnight for the launch of Windows 95. This week sees the twenty fifth anniversary of the official launch of Microsoft Windows 95.

Steve Jobs had not been at Apple for ten years so this was a heady time for Microsoft. Money was no object. I remember attending a Microsoft conference in Los Angeles during those profitable Microsoft days and one afternoon at the conference hundreds of buses turned up and took us to a surprise venue. Microsoft had hired out Universal Studios just for conference attendees. The general public had been excluded and every ride was available for free for us. The expense must have been incredible.

And so it was with the launch – dare I say it the launch that started a trend that continues to this day of product announcements being events in their own right. The entire launch of Windows 95 cost Microsoft US$300 million (in 1995 terms). They bought the rights to the Rolling Stones 'Start Me Up' and hired celebrities like Jay Leno to be live at the launch. The biggest sitcom in the world at the time was Friends so they hired Jennifer Aniston and Matthew Perry for a Windows 95 tutorial video.

But the important question: Did the product live up to the hype?

40 million sales in the first year would seem to suggest that customers could see the benefits of Windows 95 - many of which we still use today.

Windows 95 was the first desktop operating system built with 32-bit pre-emptive multitasking architecture which increased throughput dramatically when run on the latest processor available at the time - the Intel Pentium CPU!

It may seem like a childhood myth for many but before Windows 95 all filenames had to be in an 8.3 file format. For example, I may have saved this article as tktlk213.doc to give me some vague indication of the contents of the file. Looking back many months later it was often difficult to remember what was in a file when the name was limited to eight characters.

When we plug a new device in to our computer now there is an expectation that it will be discovered by the operating system and just start working. Windows 95 introduced the concept of Plug and Play – or as I often used to call the early unreliable versions, Plug and Pray. Visually, the Start button and the Taskbar were both introduced in Windows 95 and live on in modified forms to this day.

Probably the endearing memory people have is what was affectionately known as The Microsoft Sound. Brian Eno, a pioneer in ambient music, was commissioned to write a piece of music that was inspiring, universal, optimistic, futuristic, sentimental, emotional and make it no longer than three seconds! Brian delivered his little jewel with an extended play time of six seconds but his dirty little secret was that he created it on a Mac!

With all of this expenditure on Windows 95, how long was the line outside my business when we opened at midnight? Well, not quite a line. We had two customers during our one hour opening timeslot. It seems the excitement had not filtered through to our clients!

Tell me your favourite memory of Windows 95 at ask@techtalk.digital. TECH TALK 213: 10 JAN 2020

Aviation Variation Elation

Look! Up in the sky. It's a bird! It's a plane! It's...sorry, it's not Superman. As much as many people like the idea of a guy flying around with his underpants over his pants protecting us, it is much more likely that if you look up in the sky today, you will see a drone.

With a third company recently cleared by the Federal Aviation Administration (FAA) for commercial drone deliveries, we continue to edge closer to the vision of a thirty-minute turnaround from order to delivery in metropolitan areas.

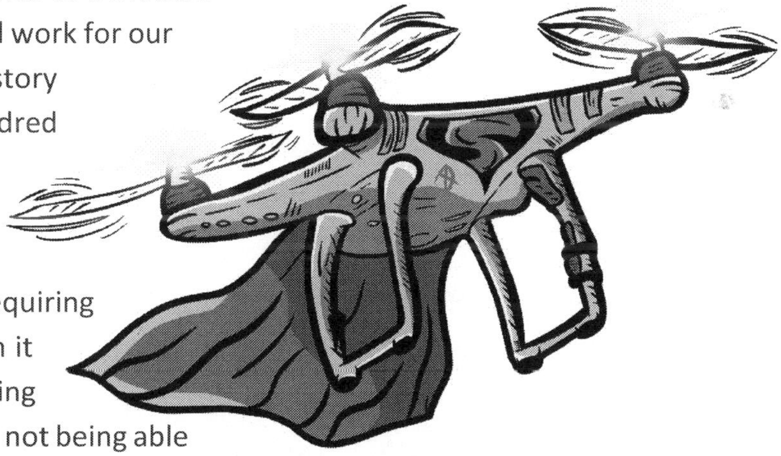

Although we see drones as a modern invention, the ground work for our current crop has a history of over one hundred years. The first quadcopter was created way back in 1907. Apart from requiring four people to launch it to its maximum flying height of a metre and not being able to steer it, it was a huge success! Modern inventions need to start somewhere.

By 1917, the first remote controlled (RC) winged aircraft took flight. It was based on RC technology from Nikola Tesla and the aim was to use the device as a flying bomb. Despite never being used in a combat scenario, it paved the way for future military projects.

The transistor was invented in 1947 which was the breakthrough needed for a multitude of electronic products. By the sixties, transistors were being used in RC planes allowing an entire industry to be created that continues on today.

When most people think of a modern drone, they think of a wingless vertical take-off device controlled via their smartphone. That format was pioneered by the French company, Parrot, who released their Parrot AR Drone in 2010. It sold over half a million units. By 2016 we had drones with smart computer vision using machine learning technology to allow them to avoid obstacles and intelligently track and photograph moving objects. DJI's Phantom 4 was the first to market with this technology – and as impressive as this technology is, I still managed to crash my first Phantom 4 drone into a tree within about ten minutes of my first flight! Ouch!

The recreational drone market is really taking off (couldn't resist sorry) with approximately four million drone sales last year and double-digit growth but the really interesting space is with the variety of commercial applications.

You may remember the world's first pizza delivery that occurred in New Zealand in 2016. Quick access to fresh food is a great practical application of a commercial drone. Various companies are now delivering coffee and donuts and burgers and even beers! Medical goods and blood supplies and incredibly body parts are other sensible uses of the convenience of the air compared to the congestion of the road. Agriculture is taking advantage of the labour saving and technology tools offered by drones with winemakers to rice growers using tools such as infrared cameras and crop yield analysis algorithms.

Technology is even being used to deliver technology with Internet access drones being trialled in areas without Internet access! The list goes on with development of drones to be used for window cleaning; wildlife conservation; construction inspections; ambulance services...you name it and it is probably being developed.

As battery development continues to charge ahead (two puns for the price of one today) and wireless technologies such as 5G improve, expect even more uses for drone technology. On the negative side, drug cartels and governments are creating more effective weaponised drones which I hope to never see when I look up in the sky. Tell me your best idea for the use of a drone at ask@techtalk.digital.

TECH TALK 214: 4 SEP 2020

Compare Share Glare

In raising our four children, my wife and I hope we have managed to achieve many things but, at the very least, we hope we have taught them to have a sprinkling of scepticism about the world around them. When something seems just way too good to be true then, well, there is a pretty good chance that it is. You know the type of thing...the five-million-pound lottery win for a draw you didn't enter; the windfall from the massive estate from the overseas deceased unknown relative and those claims of receiving something for nothing – like free energy from the sun.

Hold on. That last one might actually have something in it.

Since 1953 SW Hart and Co. (later Solahart) has been promising to give you hot water free from the sun. Moving from hot water to electricity presented different challenges and although the photovoltaic effect was first discovered in 1839, it took until 1954 just to reach 4 per cent efficiency. Efficiency has now improved to the point that a new record was set five months ago at 47.1 per cent but the efficiency of most solar panels installed on rooftops around Australia is in the low to mid twenty per cent range.

Last decade saw the start of a dramatic increase in the total amount of power generated by rooftop power. A combination of more efficient panels and a drop in the cost of the panels was helped along with some generous feed-in tariffs. The result? 11.8GW of power is currently being produced by rooftop power. We are still only scratching the surface though. A recent report had the total potential rooftop power output for Australia at 179GW (enough for our total electricity needs) but that would mean every vaguely horizontal surface on every building would be covered with solar panels. It probably isn't going to happen.

The most exciting current development is with solar glass. The major challenge for scientists is to maximise the ability for the glass to produce power whilst not compromising transparency. A new record was recently set by University of Michigan researchers where they achieved 8.1 per cent efficiency with the generation of power and 43.3 per cent transparency.

Even though the solar generation seems low compared to the panels we have on our rooftops, the huge advantage with solar glass is in surface area. Take the example of the Willis Tower in Chicago. At approximately 70m by 70m at the base and a height of 442m, the absolute maximum roof space would only be just under 5,000 square metres but just the surface area of one face would be almost 31,000 square metres. Even if the glass was only one third as efficient as solar panels, the surface area being six times greater would give you a better generation of electricity. More importantly, the electricity is being generated where it is being used. Could we even dream of a skyscraper with enough solar glass that it would be energy positive?

It comes with another positive. I remember standing on the balcony of the Mandalay Bay in Las Vegas marvelling at the almost mirror like surface of the entire building. I was told by locals that, with temperatures as high as 47 degrees Celsius, the mirrors were designed to reduce the amount of heat entering the building.

Solar glass limits the heat but also generates electricity. I think we are seeing the next step forward in materials used for skyscrapers.

Tell me what year you think Australia will produce more than 50 per cent of its power by renewables at ask@techtalk.digital. Tech Talk 215: 11 Sep 2020

NEAT HEAT FEAT

In September 1797, Lieutenant John Shortland went in the direction of Port Stephens searching for runaway convicts. He stumbled across what he later described as "a very fine coal river" and, in one of those ultimate moments of praising your boss, he took it upon himself to name the estuary the Hunter River after Captain John Hunter who was Governor of the British colony in New South Wales.

With Shortland's description of the deep-water port and the abundant coal in the area, it didn't take long before coal mining started which officially became the colony's first export. As coal-diggers and timber-cutters started settling in the area, the new settlement was eventually named Newcastle after England's famous coal port.

> By 1831 the first railway to carry export coal was opened and today Newcastle is the largest coal exporting harbour in the world.

With such a history in coal, there is a huge amount of irony in the fact that Newcastle University is doing its absolute best to shut down the use of coal in power stations across the world.

Before I explain further on that, we should look at how a coal-fired power station produces power. The concept is relatively simple. Heat up water. Produce steam. Direct that steam past turbines to spin them. By application of Faraday's Law of Induction, discovered in 1831, the moving turbines can induce voltage in an electrical circuit. After construction, the major impact – financially and environmentally – is burning coal to heat up the water. But what if we could heat up the water using a different method? Sure, burning gas or biomass are methods being used but they still require the burning of 'stuff' so they aren't perfect solutions.

Newcastle Uni has invented thermal bricks using their patented Miscibility Gaps Alloy (MGA). One argument against renewable power is that power is not produced when the sun isn't shining and the wind isn't blowing. Batteries are expensive and don't hold power forever and they require significant additional infrastructure. Storing heat in MGA is about to change all of that. There are approximately 2,425 coal-fired power plants in the world with an operating capacity of 2,000 gigawatts (GW). The infrastructure is in place already with these power stations and they are already connected to the grid.

The magic of MGA is that it will allow heat to be generated when convenient (when the sun *is* shining and the wind *is* blowing) and capture that heat within the magical block. Then when power is needed, the heat can be released to produce power. Better still? The cost of MGA is one tenth the cost of equivalent lithium batteries.

The aim from Newcastle Uni is to bridge the gap between cheap and abundant renewable energy, which is often generated in peaks, and the ability to store and dispatch power at any time of the day.

Already the use of coal worldwide is decreasing but the rate may not be fast enough with only a three per cent drop over the last year. MGA could accelerate this reduction.

The only bad news from this exciting development? Although Aussie ingenuity created this product, as so often happens with great Aussie ideas, commercialisation may rely on overseas partners. A Swiss company is the first to come on board to produce blocks for trial in a 55-megawatt (MW) demo plant.

Will MGA be the panacea the world needs to reduce our reliance on coal? Possibly, but it is heartening to know that Australian scientists are working towards innovative solutions.

Tell me if you have a better idea than MGA thermal bricks at ask@techtalk.digital.
TECH TALK 216: 18 SEP 2020

Express Excess Distress

Before COVID-19 hit the world, there were six aged care facilities in the region that I would regularly visit – sometimes with one of my children - to deliver a presentation on technology or perform poetry for the residents. On some occasions I might be interrupted during my presentation by one of the residents. I dismissed it as a resident being a bit cantankerous. Sometimes my kids would comment that a particular person seemed rude. Other residents would tell me the person in question could be a little 'difficult'.

Well it turns out that the behaviour of some residents may not be the result of impoliteness but the result of pain. And the source of this information? Artificial Intelligence (AI)!

Now I understand that there are times when AI gets a bad rap. AI conjures up images of the *Terminator* franchise or *The Matrix* where machines are taking over the world and humans are a minor annoyance. The incredibly popular *Westworld* is a continuing story of the evolution of AI in control.

As the brilliant physicist Stephen Hawking once commented, the emergence of AI could be the "worst event in the history of our civilisation" so you can understand why there is some apprehension in the community.

> AI taking away our jobs is one thing – but the destruction of mankind? That seems a little more serious.

While we wait for Rehoboam to control our lives, there are huge advantages in AI.

Back to those aged care residents. New technology is currently under development that will allow micro-expressions on a person's face to identify the presence and severity of pain. After 180,000 assessments so far across 16,000 residents, the results show that up to 80 per cent of aged care residents are suffering from chronic pain. With 53 per cent of those same residents suffering from dementia and many unable to adequately express themselves, this pain is often undiagnosed. Behaviour of residents that is often dismissed as argumentative or even aggressive is often just a person in pain.

And the best hope for these residents? Artificial Intelligence!

Now I understand that not all of us are in an aged care facility – although I hope many of us make it that far – but AI is working away in many other parts of our life. When we open up the main screen of our favourite streaming service, AI analyses billions of records and matches them against what we have viewed before, including the viewing time of day, and makes recommendations based around all of that information. 80 per cent of what we watch is driven by an AI recommendation.

Music streaming services are possibly even a step ahead. Based on several hundred musical characteristics, songs are analysed and classified and recommended based on our listening history and, seemingly by magic, we are listening to songs that we inherently love!

When I was organising a charity bicycle ride last year, my credit card was automatically blocked by my bank's AI. Multiple transactions at different service stations, supermarkets and motels over a series of days were flagged as 'unusual' and it was assumed my card was stolen. In this case the transactions were legitimate but it was comforting to know that AI was watching.

Lastly, there is a pretty good chance that the level of comfort of our retirement will depend on the success of AI as many Superannuation funds are entirely managed by AI systems.

It is both a scary and exciting world we are currently navigating – but AI is a part of that future roadmap. Tell me your greatest hope for AI at ask@techtalk.digital.

Tech Talk 217: 25 Sep 2020

Superfast Broadcast Forecast

Futurist. It conjures up the image of a person sitting in a darkened room with two hands placed on a large orb floating over a table with an unsuspecting victim waiting to hear if they should move to Toledo. In this context, possibly a better term might be charlatan.

The real futurist spends a significant amount of time researching the past, looking at historical movements, investigating breakthroughs and combining all the available data to build some form of roadmap to where a particular industry might be at some point in the future. More legwork than guesswork.

To a certain extent you would hope that the leaders in society are all futurists because not only do they need to know where we are going but they can also help shape that future.

The latest changes in the National Broadband Network (NBN) show the stark contrast between futurist and populist. Communications Minister at the time, Stephen Conroy, announced the concept of the NBN as part of an election campaign and, once elected, established NBN Co on 9 April 2009. I had my first meeting with Conroy on 17 December 2009. I could see the economic and social advantages of the NBN for our city and wanted our community to be at the forefront of regional communities adapting to the new digital age.

Unfortunately, along the way of the futurist vision, the nation had a change of government and cute three-word slogans and false economies got in the way of the rollout of the modern equivalent of the overland Telegraph line.

Our visionary future of a fibre-connected nation ended up with a multi-technology mix – or as I liked to call it a multi-technology mix-up!

A country that was rich in resources had the opportunity to lead the world in the new digital revolution and set the nation up for decades. Sure, rolling out fibre to a large nation was expensive but making do with an inferior solution that everyone apart from the government could see would need to be fixed in the future would be more expensive in the long-term.

Our current position in the world is embarrassing. Let me drop a few stats. The Speedtest Global Index currently has Australia at 68th spot. New Zealand, by contrast, sits at 24th. Out of 35 OECD countries, Australia ranks 32nd. The global average download speed is 73.5 megabits per second (Mbps). Australia's fixed broadband speed averages 41.7 Mbps.

When a former Prime Minister made the grand statement that "25 megabits per second is going to be more than enough for the average household" he missed a significant point. Not only is this way below where the world sits, the next frontier is Gigabit. In small numbers today but that is accelerating quickly.

Iceland leads the world with Gigabit test penetration of 0.71 per cent. Singapore is 0.47 per cent. Even New Zealand is at 0.13 per cent. And Australia, accurate to two decimal places, is at 0.00 per cent.

The good news is that the government has finally realised that we are being left behind and NBN has announced a roadmap to upgrade to FTTP and therefore to Gigabit speeds – and beyond. We have some catching up to do though – New Zealand already has 10 Gbps speeds. At the moment only 18 per cent of Australian premises can access Gigabit speeds but the NBN pricing has precluded most retailers from offering reasonable plans.

With the $3.5 billion upgrade Australia can finally start clawing its way back. Tell me if you would connect to Gigabit if the plans were available for you now at ask@techtalk.digital. TECH TALK 218: 2 OCT 2020

Excess Process Mess

I don't sing. I am not against songs or music but I believe it is important you know your limitations. Holding a tune is not a talent I have so I don't wish to subject people around me to listening to terribly out of tune singing.

Similarly with technology. It is important to know limitations. No matter how advanced the technology, there are always limits. They may be high limits or limits we never dream of reaching, but there are limits nonetheless.

It was many years ago that some people considered 640KB of RAM would be enough for a personal computer. When 32-bit operating systems were the norm, the addressable memory was 4GiB which was considered a huge amount of memory. Now, with 64-bit operating systems commonplace, the addressable memory space is four billion times larger at 18EiB which seems way larger than we would ever need…but it is a limitation nonetheless.

Once upon a time, people as clever as Aristotle believed that light was instantaneous. When the Dutch scientist, Isaac Beeckman tried to measure the speed of light in 1629, he used gunpowder and mirrors but the distances were too small. Danish astronomer, Olaus Roemer, finally measured the speed of light with some accuracy in 1676 using the moons of Jupiter as a reference point. We now know the finite speed of light as 299,792 kilometres per second - which brings us to another limitation. We think of our communication network – voice and Internet and broadcast services – as instantaneous and, for the limitation of humans they typically are but…if you introduce geostationary satellites in to the mix, the latency can be noticed by humans.

At 35,786 kilometres above the earth, using a satellite for Internet access gives a 477-millisecond inherent delay with every action.

Mostly we know and understand these limitations and work within them but when we don't know our limits, much like if I was to sing in front of an audience, the results can be disastrous.

It is well known that the National Health Service (NHS) in England lost a significant number of records of COVID-19 impacted residents. The cause of the loss has now been revealed. The lack of understanding of the limits of Microsoft Excel. More specifically the limits of the old data format.

Excel was first launched in 1985 with the Windows version launched in 1987 which had the file format of .XLS. The total size of one spreadsheet was limited to 65,536 rows by 256 columns. A major upgrade occurred in 2007 which increased these limits to 1,048,576 rows by 16,384 columns. The new file format was .XLSX.

Unfortunately, someone at the NHS wasn't aware of the significance of that last X in the file format and used a pre-2007 spreadsheet to collate the data logged from COVID-19 tracing. The initial logs were collected and the data was fed in to an Excel template. Once the information reached the row limit, the additional cases were simply truncated.

Using the new file format would have increased the number of cases by a factor of sixteen before cases started dropping off which indicates that Excel – no matter the version – may not be the best product to use for contact tracing. A SQL database, for example, has a limit of 524,272TiB per database with the number of rows only limited by the size of the total database. That calculates out to…a big number. In technology, much like in the rest of the world, knowing the correct tool for the job is critical.

Let me know when a technology limitation has impacted your life at ask@techtalk.digital. Tech Talk 219: 9 Oct 2020

Hello Studio Photo

A moth navigates by transverse orientation. It keeps the dominant light source at a certain position in relation to their body to guide them. When we turn on a light on our verandah, those same moths keep flying around the light until they finally meet their demise. As much as they try and resist, the decoy of the light is too much.

In a similar way I was up at 4am on Wednesday this week. As much as I wanted to sleep and read about the event in the morning, the lure of yet another Apple launch event was too much for me. Try as I may to resist it, I sat in front of the television watching the event and writing notes for this column.

I saw four new iPhones with two 'Pro' models and two 'Standard' models. In the Steve Jobs era, Apple events launched revolutionary products.

> Apple launches now are a mix of catch-up plus evolution with possibly a dash of revolution.

Firstly, Apple announced that all four of their new iPhones would be 5G. The way Apple talked about 5G you would think that, if they were not the ones that invented it, they were certainly the first to market with it. 5G phones have been available since April last year. This 5G would be better than the 5G on other handsets Apple continued. It covers the sub-6GHz frequencies in addition to some of the mmWave frequencies to make the phone able to be used in more locations around the world. Unfortunately, only the US version of the iPhone covers the 28GHz and 39GHz frequencies. The phones sold in Australia and other markets only support up to 4.7GHz.

The next major battleground is in cameras. Surprisingly, the size of the sensor and quality of the lens are not the only critical components in photography. The processing power is of utmost importance. The new A14 Bionic chip packs in 40 per cent more processors than the previous chip. This increase in power allows better photography, particularly with Night Mode. The two Standard model iPhones have two rear cameras and one front facing camera. The two Pro models though sharpen the focus for serious photographers. Three cameras including a telephoto lens with a 4x optical zoom range (5x for the Pro Max) and the introduction of Apple ProRAW letting the user manipulate the RAW images shows that Apple is serious about imagery.

Across the four iPhones, screen sizes went up and down. The iPhone 12 Mini is the smallest Face ID iPhone ever at 5.4". The iPhone 12 and Pro both measure 6.1" followed by the largest ever screen on an iPhone – the 6.7" Pro Max. All that extra screen space sounds like a cracked screen waiting to happen but the new Ceramic Shield on the iPhone has four times the break resistance of the previous material.

There are many other evolutionary enhancements from MagSafe to squarer styling reminiscent of the iPhone 4.

Now for the revolution. This is not an addition but a removal.

With two billion Apple charging bricks across the world and seven hundred million Lightning Port headphones, Apple has chosen to leave both out of the packaging to reduce e-waste. While there is logic in this approach, and I expect others to follow, the charging cable included with the iPhone is Lightning to USB-C. Unfortunately, the majority of those two billion charging bricks are USB-A. I can see this 'environmental' step introducing additional revenue for Apple!

Let me know if Apple has done enough with this upgrade to entice you to upgrade at ask@techtalk.digital. TECH TALK 220: 16 OCT 2020

Tune Moon Soon

Amenity. The main driver for deciding where to live is that all-encompassing term 'amenity' which can really mean what you want it to mean. I looked at a population heat map in Australia from the turn of last century and the country was smattered with many small dots as people lived where they worked.

Then two major disruptions occurred.

Firstly, in 1908, Henry Ford had the first of fifteen million Model T Fords roll off the production line. Colonel Harley Tarrant assumed the responsibility for distribution of Ford automobiles in Australia in the same year. Fast forward fourteen years. Alexander Kennedy became the first Qantas passenger on a scheduled flight.

As efficient means of transport were developed, people could live further away from where they worked and that esoteric quality of amenity started to become more prevalent in the decision-making process. A current population heat map of Australia shows fewer population centres with larger populations.

In amongst amenity, there are many factors to consider. Road access; electricity; water; sewerage access and more. Many people would consider electricity essential and I would say that an increasing majority would consider the NBN in that same category.

What about mobile phone coverage? Personally, I would say yes.

Well it is such a strong factor in the amenity factor that NASA has just contracted Nokia to build a 4G mobile network on...the moon! The network will be up and running by the end of 2022.

It does sound crazy that you will have reception on the moon but your call drops when you walk in to a building here on earth. Maybe the astronauts dug their heels in and refused to go to the moon unless they had a mobile network! "The amenity just isn't quite right," may have been their argument.

The 4G network won't just be used for an astronaut to call home and read a bedtime story to their child – although that might be interesting with the 384,402km distance to the moon creating some stilted conversations with the 1.28 seconds of latency. In addition to voice and data, the mobile network will also be used for navigation; streaming biometric data back to earth; remote controlling robots and checking sensors.

The 4G network on earth is supported by large cell towers with reliable power or generators. One advantage on the moon is the lack of trees and buildings to interfere with the signal. The disadvantage is that you have to transport the equipment to the moon. Nokia will be using small cell technology that uses less power than traditional cell towers and is easier to fit in to a rocket.

This is all part of the NASA Artemis program. After the mobile network is operational in 2022, the first woman and the next man to walk on the moon will arrive before the end of 2024 and by 2028 there will be a sustainable base on the lunar surface. NASA has already committed to deals with fifteen companies worth over US$370 million to establish the base on the moon.

And the next big step? Travel 65.9 million kilometres to setup base on Mars. I can only assume that there will be at least a 5G network on Mars before that time! The next frontier truly is space.

It's one small step for phones but one giant leap for phonekind. Very exciting news for moonwalkers and possibly just a bit annoying for anyone who can't get a signal right here on Earth.

Tell me if the mobile phone network is an essential component of your desired amenity for where you live at ask@techtalk.digital. Tech Talk 221: 23 Oct 2020

Lane Maintain Pain

Do you remember the original Pixar/Disney Cars movie? It was launched back in 2006 and was somewhat successful. It grossed $462 million at the box office but also generated $10 billion in merchandise sales in the five years after its release – both figures which I contributed to!

In one scene, *Doc* introduces us to *Bessie*, the "finest road paving machine ever built!" *Bessie* was towed behind *Lightning McQueen* to lay asphalt on a road.

The idea that a piece of equipment as simplistic as *Bessie*, with her furnace, asphalt container and heavy roller, could build a road appealed to the target audience but is probably an insult to road maintenance workers across the world. I attended the National Local Roads and Transport Congress (more exciting than it sounds) in Mount Gambier in 2011 and after one session that explored all the engineering work that was required for the building of a road, I asked the clever engineers if a machine similar to *Bessie* would one day be invented to make the job easier.

Laughter was my answer!

Well, laugh no more. Liverpool, famous for rock music and soccer, may soon be famous for an autonomous road-repair robot! Researchers at the University of Liverpool have developed an Autonomous Road Repair System (ARRES) that will autonomously patrol streets, assess the condition of the roads and fix cracks and potholes as soon as they are detected. Road closures for repairs will be a thing of the past as the vehicle will stop, notify surrounding traffic of its presence and complete the repair within a few minutes. Logically it would operate through the middle of the night in the more congested areas and it can operate for many hours at a time between recharges.

But it gets better.

Rather than rely on ARRES to perform all of the survey work, the road condition assessment technology can be integrated on to existing vehicles that are already travelling over many parts of a community – such as garbage trucks and taxis. While these vehicles go about their normal activities, they would be collecting data for ARRES to then decide where to travel to next. Once there, a repair would be completed and ARRES would move to the next problem area.

Prevention is certainly better than cure and this system would assess the road condition and repair areas that show small cracks and deviations before they became major issues. Saving money on repairs, reducing repairs on vehicles travelling over those roads and ultimately saving lives.

I spend most of my time looking well in to the future but the road damage detection technology will be on the market within six months and ARRES itself will be up and running by the end of next year.

This is all part of the changing landscape we see as four crucial technology areas converge. The increased computing power that can deliver Artificial Intelligence solutions. The rapidly improving area of robotics and what a modern robot can achieve. The range and recharging time a modern battery can deliver. And the ubiquitous connections and real-time data that we have access to with our mobile phone network and GPS systems. Remove any one of those four aspects and the creation of ARRES would be difficult, if not impossible.

So ARRES is not quite the magical machine that I would like to see building our roads, but this is just step one. To those engineers who laughed at me back in 2011, give it another decade and I will have the last laugh!

Tell me the first road you would put ARRES on at ask@techtalk.digital. Tech Talk 222: 30 Oct 2020

Rare Bare Share

I try not to stand naked in front of strangers. Actually, I try and avoid doing it in front of friends as well! The problem is, that fear of being seen in the nude is impacting our health – particularly with men. Only 40 per cent of men over the age of 35 visit a doctor regularly and one of the primary reasons they state for not visiting a medical professional is embarrassment. Embarrassment about their body, about the state of their health and about the shape and size of various bits and pieces.

In Australia that can create even more of an issue. Over 67 per cent of Australians will be diagnosed with skin cancer before the age of 70. We officially have the highest rate of melanomas in the world with New Zealand coming in second (at least we can beat New Zealand at something) and men have a higher rate than women.

The good news with skin cancer is that we have methods of detection and we have methods of treatment when detected early – but it involves being naked in front of a doctor. Or does it?

Two crucial statistics helped form an idea for developers to create a new app. COVID-19 has resulted in a 94 per cent drop in medical visitation for general check-ups. Males were reluctant to visit a doctor before coronavirus but now they had a somewhat legitimate excuse. Social distancing. The second statistic is that 32 per cent of men over the age of 35 are quite happy taking a photo of parts of their naked body and sending it to another person.

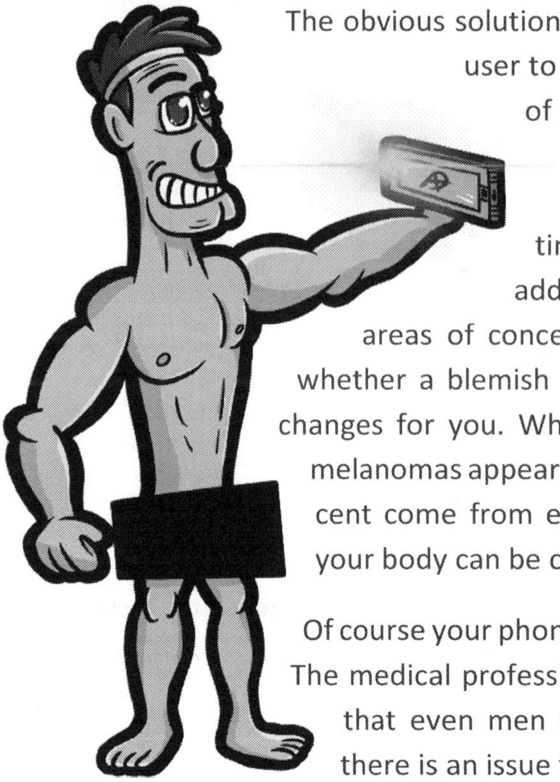

The obvious solution is to develop an app that will allow a user to take a naked photo and let the power of AI use body skin mapping and augmented reality to digitally track freckles, moles and skin changes over time. Users take full body photos in addition to close-up photos of specific areas of concern on their body. If you aren't sure whether a blemish has been changing, let the app track changes for you. When you consider that 80 per cent of melanomas appear as new moles or marks and only 20 per cent come from existing moles, recognising changes in your body can be crucial to catching a melanoma early.

Of course your phone can't resolve an issue once detected. The medical profession is still needed but research shows that even men will visit a doctor once they confirm there is an issue to be dealt with.

But this is only the start. There is a new mobile health category that is growing at 40 per cent each year and it will be worth US$22 billion by 2022. Although the trend was already happening, there is nothing like a pandemic to accelerate the progression of this market.

Researchers at the Massachusetts Institute of Technology (MIT) have developed an algorithm that will correctly identify a person infected with COVID-19 just by the sound of their cough. What is even more remarkable is that it will pick up the difference even in people who are asymptomatic. Although this hasn't been turned in to an app…yet…imagine the practical examples of using such an app for daily screening of employees or students. Gaining entry in to an event just by performing a cough test. After collecting 70,000 audio samples the researchers have refined the accuracy to a level of 98.5 percent. Not bad considering that saliva tests are only rated at about 70 per cent.

Tell me if you are comfortable with AI powering some of your health checks at ask@techtalk.digital. Tech Talk 223: 6 Nov 2020

Decode Barcode Mode

Back in 2015 I thought I was being very clever. Handing out a 90mm x 54mm card with my details on it seemed a little old-fashioned in a modern digital age...but many people still believed business cards were the single greatest marketing tool available. To try and modernise the concept, I printed a QR (Quick Response) code on the back of my cards (complete with part of our logo embedded). The idea was that I could hand my business card to any person with a smartphone and they could simply use the QR code to correctly enter all my details in to their Contacts list.

> Sounds incredible...but before March this year, the sum total of people who actually used my QR code was approximately one. Me!

Despite the fact that QR codes had been around since 1994, when I turned my card over to show people what was on the back, I normally received a blank stare. Maybe they just didn't want my details?

COVID-19 changed all of that. With contact tracing at cafés and QR codes compulsory from 23 November 2020 in NSW at all hospitality venues, the public is certainly more familiar with QR codes. I thought I should explore the technology behind the matrix of black and white squares.

Before the QR code was the barcode. The barcode was patented in 1951, based on the concept of Morse Code, but was initially an invention looking for a commercial use. In 1974 the first Universal Product Code (UPC) was scanned on a pack of Wrigley Chewing Gum. Before this time, a checkout employee would enter the amount of each item in to a cash register and it would sum the cost of the items. All stock control was a manual process and errors were easily made. The UPC allowed a unique 12-digit number to be assigned to all products (12 digits allows one trillion unique barcodes) and the automation of the supermarket checkout process had begun. To understand what occurs when a barcode is scanned, it effectively simulates a human typing in 12 digits on a keyboard and pressing Enter.

The main limitation is the amount of information able to be stored. Some barcodes only allow numeric data while others allow alphanumeric. Some barcodes allow a maximum length of 80 characters but more commonly barcodes represent just 12 and they are typically numeric. If all you want to do is record and read 12 numeric characters, a barcode is for you.

With the realisation that there are times when you need significantly more information, the QR code offered significantly increased density. With a maximum of 4,296 alphanumeric characters, it gives significantly more scope to allow the easy entering of detailed information.

The QR code was initially used to track vehicles and components during the manufacturing process. While some people have expressed concern about privacy issues and a fear of QR codes when they enter their local café, the QR code itself is harmless. It is effectively a string of letters and numbers written in a format that a computer can easily translate. When you walk in to a café, the owner could just as easily have a Web site address that you need to manually type in to enter your details for the purpose of contact tracing. The Web site address is not to be feared but if you have concerns it would be what occurs with your information after you enter it in to a Web site or an app.

Tell me if you think people will now start scanning the back of my business card at ask@techtalk.digital. TECH TALK 224: 13 NOV 2020

Nose Shows Dispose

My daughter came to me this morning and asked if she could use the milk in the fridge on her porridge. "The expiry date was yesterday," she said. "Is it safe to use?" I did as any caring parent who wanted to avoid an early morning trip to the supermarket would do, and answered with "they always build in a safety margin so it will be right to drink" hoping that I was correct and wasn't about to subject my daughter to some horrible illness.

The use by and best before dates printed on foods are the responsibility of the food supplier. There are so many variables with how the food is stored and how well it is sealed after opening that the dates printed on the packaging are naturally very conservative.

That conservative nature leads to a significant amount of food waste.

What we really need is a definitive answer. Is this particular food item safe to eat?

It is not surprising that researchers have been working on exactly this issue, and a group of scientists from Singapore, in collaboration with teams from China and Australia, have developed an e-nose specifically for the decay of various meats.

The e-nose relies on 2 components. Firstly, a twenty-segment barcode is placed inside the packaging of the meat. As the meat decays, the various lines on the barcode change colour in reaction to the gases produced. The concept was based on trying to mimic our noses. When we smell meat that is decaying, the odours bind to receptors in our nose which generate specific signals that are transmitted to our brain to tell us that meat is 'off'.

In the e-nose the bars on the barcode act as the receptors. Each bar is made of a natural sugar embedded on cellulose and loaded with a different type of dye. As the type and concentration of gases are emitted, the various bars change colour. The unique combination of colours serves as a 'scent fingerprint' to indicate the state of decay of the meat.

It would be possible to have a chart that outlined colour combinations of the barcode that you could use as a reference to determine the state of your meat. But that would be clumsy at best.

Enter the second crucial component. Your smartphone.

> By using the app developed by the researchers you simply point your phone's camera at the barcode and it will give you an accurate reading of the decay level of your meat.

The app is powered by Artificial Intelligence (AI) and initially chooses from a large library of barcode colours but is also continuing to learn more as it is exposed to more tests. The initial work involved 3,475 barcode images and, just with that sample size, the accuracy of the system was rated at 98.5 per cent. As more barcodes are read from more samples, that accuracy will increase.

The ultimate aim? Your daughter working out the freshness of food? Sure, that is one thing, but the real aim is to reduce food waste across the world. In Australia alone, we waste over seven million tonnes of food each year which costs the economy $20 billion and accounts for more than five per cent of Australia's greenhouse gas emissions. Not all of this waste is in meat which is what the e-nose is specifically focused on. But it is a start – and I am sure we will see other technological advances in relation to this ongoing problem.

Tell me if you would trust an e-nose to tell you if food was safe to eat at ask@techtalk.digital. TECH TALK 225: 20 NOV 2020

CREATE UPDATE INNOVATE

When we purchase products, we should be thankful for the Innovators. Everett Rogers released his Diffusion of Innovations book in 1962 but his hypothesis is still used today for the methodology of adoption of any new idea, in particular technology.

Rogers postulated that only 2.5 per cent of society can be classified as Innovators in the sense of how new ideas are adopted. Innovators are willing to take risks, rely on scientific information, are very social and are willing to adopt technologies that may ultimately fail. Innovators will be the first customers to try a new product which is usually the time those products are the most expensive. The four other categories are: Early Adopters; Early Majority; Late Majority and Laggards at 13.5; 34; 34 and 16 per cent respectively.

The cost of technology can often be very expensive at release date and then declines over time as production increases. When I installed my 10kW of solar panels almost a decade ago, it added 5 per cent to the value of my house. If I installed the same array today, it would be lucky to make up 1 per cent of my house value.

When a technology company is planning a release of any new product, they fully understand that they first must sell to the Innovators. To access the other 97.5 per cent of the market, the Innovators need to be on board. The advantage for the 97.5 per cent – and the reason I said we should be thankful for the Innovators – is that the Innovators will pay the highest price and do the real-world testing before the rest of the market starts to buy the product. They want the latest now and they will work out the cost later. Sure – there are times that the products are not perfect and this group can live with that.

By the time the Early Majority starts purchasing the product, they are comfortable it is fit for purpose and the pricing may even have dropped.

Which brings me to electric vehicles (EVs).

Only 0.6 per cent of all new cars purchased in Australia last year were electric. Pricing is coming down; range is being tested and more charging stations are being built. The Innovators are driving (pun time!) the way forward for the rest of society and not only will the rest of society be beneficiaries but so will the planet.

Due to the higher purchase cost of an EV, current EV owners are paying higher Stamp Duty; GST and Luxury Car Tax. I have previously written articles on the Total Cost of Ownership of EVs being similar to ICEVs with the higher purchase price offset by the lower running costs. From a tax perspective, EV owners have paid their tax upfront compared to ICEV owners paying their tax with fuel excise (which, by the way, hasn't been linked to road funding since 1959).

Unfortunately, in this country, we are now seeing some Laggard Governments deciding that taxing EVs for each kilometre driven is a good way forward. When other locations around the world are forging ahead with percentage of EV sales (Norway; Netherlands; California and China are at 55.9; 15.1; 7.7 and 4.9 per cent) the last thing we need is a barrier for entry in to the market.

Not tomorrow, but one day in the future, we will run out of minerals to dig up from this nation and when we do, we hope Governments had a vision for the future rather than relying on the past.

Tell me where you sit on the Diffusion of Innovation scale at ask@techtalk.digital.

TECH TALK 226: 27 NOV 2020

Recruit Astute Fruit

Australia is a popular place to visit for backpackers of the world. For those of us that live here, that makes perfect sense. Australians are welcoming and generous and relaxed. Just as importantly, it is relatively easy for backpackers to secure some work, particularly as part of harvest crews on farms across the nation. At any given point in time, we typically see over 200,000 backpackers in our country.

As with so many aspects of our society at the moment, this year is a little different. With various reforms plus a little pandemic thrown in, we currently have less than 40,000 backpackers in the country and that number is decreasing.

That creates a major problem for farmers with harvest work to be done.

The answer may well be Clive. No, not some larger-than-life former politician with a grand plan for sending the unemployed metropolitan residents to regional areas. A fruit picking robot named Clive.

Although this will be an Australian first, using robotics for fruit picking is being developed across the world. At Wimbledon each year, 34 tonnes of strawberries are consumed by fans. A robot was recently unveiled that will team together in a group of fourteen such robots to take seven days to pick all the strawberries required for Wimbledon.

Not far away, at the University of Plymouth, a robot is currently being trialled that will pick 25,000 raspberries in a day. Not only does that dwarf the 15,000 that a human worker can manage in an eight-hour shift, there is no need for the robot to stop for food or toilet breaks or gossip sessions around the water cooler.

In Queensland, a three-metre-tall robot with nine arms is being tested for mango harvesting. With each arm capable of selecting, picking and depositing a mango in five seconds, it starts to make very light work of an entire mango crop.

I seem to talk about New Zealand innovations way more than I would like but a recent development out of Massey University in New Zealand is a kiwi-fruit harvester. Kiwi-fruit are relatively easy to harvest as they hang down from an overhead canopy but this unit can harvest the fruit, sort them into bins by size and deliver them to a pack house.

Back to Clive.

Clive will soon be working on two Goulburn Valley properties to see if a robot is a commercially viable alternative to humans. Sensors and cameras are used to view the fruit, decide if the fruit is the right colour and size and free of blemishes, and then use a suction arm to vacuum the apple, pear or stone fruit from the tree.

There is a strong belief that if our exceptional quality fruit is to be competitive on the world market, the way we can compete with countries that have lower labour costs is with technology.

The use of robotics always brings up issues related to the reduction in the workforce and the ethics associated with machines taking over. What happens next year if the 200,000 backpackers have no jobs in this country. We might miss out on revenue that those visitors generate in other areas. What about the issue of fault? If a robot damages fruit that is picked, who will compensate the farmer?

Technical issues as well as ethical issues will be answered as time goes forward, but there is a pretty good chance a piece of fruit you eat in the next five years will have been harvested by Clive – or one of his siblings.

Tell me what you would name a fruit harvesting robot at ask@techtalk.digital. Tech Talk 227: 4 Dec 2020

Airport Passport Support

I was very excited this week. Back in the old days (pre-pandemic) I used to visit six different aged care facilities on a monthly basis and perform poetry and talk to the residents about technology. COVID-19 put a stop to that but with the easing of restrictions this week, I was allowed to start visiting residents again.

The only requirement the facilities asked for was proof of my flu vaccination for this year.

I had a few options. I could look through my paper records to try and find the original paperwork; I could visit the pharmacy and ask for them to reproduce my paperwork or I could look on my Medicare records to see if the immunisation history was there.

I remember when our children were born. We were presented with our 'Personal Health Record' booklet, colloquially known by everyone as the 'Blue Book.' If you went to visit the doctor or to day care, you had to have the Blue Book with you that contained all of the records of immunisation and other health information. Heaven help the parent who lost their Blue Book. House on fire? Grab the kids, the Blue Book and the family photos – in that order!

There is no doubt that COVID-19 has accelerated the technological advancement across the world by more than three years. We have had the technology for many years to keep all of our health information and records electronically but it may not be the health authorities that finally drive the change. The International Air Transport Association (IATA) represents 290 airlines worldwide. They are developing an IATA Travel Pass as a way for travellers to show COVID-19 vaccinations and test results. They may well make use of such a travel pass compulsory before flying internationally.

A Digital Vaccine Passport (DVP) makes a lot of sense.

However, I would prefer my government to be in charge of it rather than an airline association. Most of the information is already there for a government to use. NSW has the Service NSW app which already has our Driver Licence and Working with Children Check details in it. Adding immunisation records would not be a major undertaking – although a national approach rather than a State based approach makes more sense. The largely unsuccessful COVIDSafe app is already run by the Australian Government and once again could be easily adapted to provide the required information. The logical process though would be to use the Medicare app. Basic immunisation information is already contained within our Medicare records and the app already exists. Adding COVID-19 information would seem like a logical step forward. Then the Australian Government could work with IATA and other international organisations to have the information in the Medicare app recognised by those organisations – making it much simpler for users.

There are valid concerns from privacy advocates and these have to be weighed up against the ability for society to return to something like normal. There is no doubt that the collection and storage of this information would need to be carefully scrutinised to ensure it wasn't misused.

The larger issue may relate to access to day-to-day services if authorities decided that only people with a DVP could enter certain locations such as churches, cafés and malls. The entire process also has the possibility of exclusion for people who don't have a smartphone. We are already seeing that with compulsory QR Codes for entry to hospitality locations in NSW but this might go a step further.

Tell me if you would be comfortable with your vaccination history on a Digital Vaccine Passport at ask@techtalk.digital. TECH TALK 228: 11 DEC 2020

Stop Shop Flop

My kids have one major issue at Christmas time. You would think it would be a time of excitement and presents and family gatherings but the pressure on my kids to get a good present for their dad is immense. When you have an individual that has technology as their only passion then buying a tech gift would seem like the obvious solution. Unfortunately for my kids they know I already have the latest cool gadgets so it makes it a bit tough.

I am not Robinson Crusoe though so I thought a Technology Gift Guide would be appreciated by any readers that have a relationship with a similarly afflicted tech head.

Number one on my list is a periodic table of the elements. It doesn't sound that exciting – we all learned the 118 elements back in high school. Except this 2cm thick acrylic masterpiece contains real elements. Only 85 of them with some notable exceptions of the radioactive variety and a few too rare or expensive to include. Makes a nice addition to the desk of any confirmed tech head.

When you hear someone complaining about the amount of time they spend mowing their lawn on the weekend, you have the perfect recipient for a robotic lawn mower. Modern versions of these autonomous grass cutters are whisper quiet and cut very close to the edges. There is a small task to perform when you initially purchase them in laying out the guide wire around the perimeter but after that, you just schedule the lawn to be mown twice a week and revel in your new found weekend time.

Do you have someone in your life that likes to be on time? I mean really on time? Well...they might need an Internet connected clock that keeps the time accurate to within 200 milliseconds. Network Time Servers are located around the world and these clocks use the Network Time Protocol to synchronise their time with a time server every ten minutes. They will even adjust for the time it takes to receive the signal from the nearest time server.

Hanging a reprint on your wall of *Salvator Mundi* by Leonardo da Vinci (or the original if you have a spare $470M) is nice, but even the most expensive painting in the world might become a little boring. The solution? Hang a modern digital frame on your wall and subscribe to a service that allows you to access a library of tens of thousands of curated artworks licensed from museums and archives. Set up a rotation on your preferred schedule and enjoy art worth billions of dollars for your small subscription fee.

The first Polaroid Instant Camera was produced in 1948 but I mostly remember the Polaroid non-folding instant cameras that only the rich kids owned when I was in high school. Imagine clicking a button and being able to see a completed photograph within minutes? That was fine...until digital photography. But...it seems that everything old is new again. A good option for the tech obsessed in your life is a Mini Photo Printer. Connect to your phone via Bluetooth and choose a photo to print. Send it to your nearby portable photo printer and watch a small print appear – complete with peel and stick backing allowing you to stick it anywhere.

Hopefully that gives you a few possibilities for presents for your friends and loved ones.

And a big sorry to my kids – I already have all of these items on the list so you will have to look harder!

Tell me your best idea for a tech present at ask@techtalk.digital. Tech Talk 229: 18 Dec 2020

ACE DATABASE TRACE

Significant dollars are spent on market research and statistical services across this nation. With over 14,000 people employed in the sector it is worth over $3 Billion (US$50 Billion worldwide). While I don't propose that we should shut down this industry, we can gain a small insight into what is happening in every home and business across the nation by looking at one piece of information. Google search data. With the notable exception of the Census, most market research takes a sample size and extrapolates the data to make educated estimations whereas Google search data gathers information from every single search performed throughout the year.

And what did we learn from looking at that data?

I am disappointed to say that the number one search term in Australia for the year was..."US election." The election only occurred on 3 November yet it was the most searched term for the year! I am not sure why we are so fascinated with the US but the data doesn't lie. Number two was no surprise. "Coronavirus." With most of the terrible bushfires at the beginning of this year extinguished by 10 February, it is somewhat surprising that "Fires near me" was number three on the list. "Coronavirus Victoria" was number four followed by "Toilet paper." I am not sure if people were searching where to buy it, how to make it or what to do with it, but we seem to have a fascination with the ten-centimetre square of two ply that hangs in every home.

When I look at the most popular sport terms for the year, we again see the fascination with the US. "NBA" was number one showing the attraction that is American basketball. English soccer is still a very big sport with "EPL" number two and "State of Origin 2020" was the first Australian sport at number three. Soccer was featured again at number four with "Champions League" and cricket rounded out the first five with "IPL" fifth.

When I researched searches that included the term "How to..." it was certainly obvious that COVID-19 made a huge impact here.

"How to make hand sanitiser" was number one – although I am not sure I would completely trust a home-made hand sanitiser.

"How to make a face mask" was number two and number three reinforced a comment I have made several times during the pandemic. We have accelerated our use of technology by about three years in the first three months of the pandemic. "How to use Zoom" was number three on the list. "How to get tested for coronavirus" was next on the list and "How to make bread" rounded out the top five. Sales of flour increased by 160 per cent over 2019 and cookbook sales went up by 15 per cent. People sitting at home suddenly had time to do things they wouldn't normally do – like bake bread. Except, looking at the Google data, they didn't actually know how to do it.

Lastly, when I pulled the Google data that included the term "Can I..." I found a list completely driven by the pandemic. "Where can I buy toilet paper" topped the list with buying hand sanitiser a close second. Finding a location to be tested for coronavirus was third and the government should be pleased with itself that the fourth term was "Can I access my super" showing that early access to superannuation was a popular move. We also missed visitors with "How many people can I have in my house" at number five.

Let me know which of these results surprised you the most at ask@techtalk.digital. TECH TALK 230: 25 DEC 2020

Quite Bright Highlight

Throughout the year I have talked technology across the airwaves with over three hundred radio segments and written over fifty technology columns. What were my technology highlights from talking and writing about technology throughout the year?

My overall highlight? Adaptability. The world as we knew it changed dramatically over a very short time frame. Technology had to be adapted to a new world. And, by association, people had to adapt. If ever we doubted the resilience and flexibility of humans, the year of COVID-19 has shown us not to doubt.

The first technology highlight was videoconferencing. The TV series '24' aired an episode in 2007 which showed the Cisco TelePresence suite. It seemed like only months before everyone would be sitting in their home office and performing a virtual commute. Well, it took thirteen years and a global pandemic but the promise has finally been delivered. With everyone from grandmothers speaking to their grandchildren down to politicians participating in parliament in a virtual sense, there are few people that haven't tried the concept at some stage this year.

In a similar vein, many people had to learn how to better use their mobile phone this year. Whether it was loading a COVID related app or to scan a QR Code to have a coffee, some mobile phone users in the past were happy just to use their phone as, well, a phone, despite the incredible power contained within. COVID-19 forced many people to discover some other uses of their mobile and be exposed to just a few of the millions of apps available.

While not exactly mainstream yet, the ability for your phone to be used as your 'key' to access your car and home was another highlight of this year. Tesla has had the ability for several years but Apple launched the Digital Car Key this year. In the not-too-distant future, we will laugh at the 'old days' when we used to carry different keys and different remotes for the variety of access devices we required. Our phone or our smartwatch or our body will allow us access in to all the places we are allowed to go.

While some car companies have been under-delivering on the promise of autonomous driving, this year was still a year where we saw significant developments in machine autonomy, somewhat aided by progress in AI. With everything from fruit-picking robots to autonomous road-repair machinery and drones and cars improving their ability to self-navigate, there is no doubt that we are edging ever closer to autonomous devices. The Federal Aviation Administration (FAA) cleared the path for commercial drone deliveries and the California Public Utilities Commission (CPUC) approved robotaxis. My kids will need a license to drive a car but I am not sure my grandkids will need to worry.

We saw significant movement in attitudes towards power production this year. Coal was still king, but when it was revealed that mining giant, Glencore, spent up to £7 million per year to use social media bots to amplify their messaging and with Climate Change further acknowledged by the public, the market has started moving more rapidly to renewable power.

Non-renewable power still made up 79.1 per cent of power production in Australia in 2019 but at current trends I predict that 2026 will be the year that we hit the 50 per cent mark – and that is without help from the government.

Increased efficiencies we witnessed this year in solar panels and solar glass as well as further technological advances in power storage will help this progress.

Tell me your 2020 technology highlights at ask@techtalk.digital. TECH TALK 231: 1 JAN 2021

Great Fate Estimate

As a technology futurist, I perform my research and gather my data and use my experience to predict what medium and long-term trends we will see in the technology space. But somewhat like a Climate Change scientist can tell you the long-term effects of global warming, but can't tell you if it is going to rain tomorrow, short term technology predictions are much harder. I am a sucker for punishment though. Each year I can't help myself. I will make some estimations of how the technology world will look just a year later and then look back in a year and see how close I was.

This year was even a little harder. On 31 December 2019, health authorities in Wuhan, China, confirmed that dozens of people were being treated with pneumonia-like symptoms. Researchers later identified a new virus and it is fair to say that the world is now way too familiar with COVID-19. On a positive note, I credit the 2020 pandemic with progressing technology three years in three months. Video-conferencing; working from home; increased Internet speeds; changes in cybercrime; additional streaming services...pleasantly, when I looked back at my predictions from last year, without knowing how our world would be upended, some of my predictions involved many of these areas of technology.

With the swagger of success under my belt, here are my predictions for our technological world in the year 2021.

Virtual Reality (VR) and Augmented Reality (AR) are often confused as the same thing – but there is a significant difference. VR takes you somewhere else and AR adds to your real-life situation. With increasing computing power and better graphics – plus a strong demand – both technologies will see a dramatic increase in usage and functionality over the next year. No – I am not just saying that because I have spent way too much time playing on my son's Oculus Quest 2 over Christmas. With restricted travel opportunities and a desire to visit people and places, what better way to do it than with a VR headset. And with physical shopping reduced, imagine placing furniture in your house in an AR scenario before ordering the physical items online and having them shipped to you.

Educators have found an incredible challenge this year. Skilled educationalists 'connect' with their audiences and ensure they are presenting relevant material in a way that is able to be absorbed. Without any additional training, those same professionals were talking to a camera through 2020 and trying to have the same impact. I expect some significant changes to occur in online and digital learning platforms and many of these will involve both VR and AR.

Oh, and there are some pretty cool games as well!

For many years I have been espousing the benefits of working from home. Work should be a thing you do rather than a location you visit. While some companies spoke about such a policy, it didn't really happen – until it had to. Employees quickly realised that commute time had disappeared and a home coffee machine was almost as good as spending hard earned cash with the local barista. Sure, some missed the office gossip but, in the hands of the correct employees, productivity could increase. To do it properly, Internet speeds need to improve (particularly upload speeds); security needs to be tightened and employers have to 'believe' but expect to see technology to allow for more working from home solutions over the next year. As an added bonus, you can use the saved commute time to play on the VR headset!

5G is not hot off the press but still new enough that conspiracy theorists like to blame it for everything from coronavirus to cancer and maybe even the lack of form of Steve Smith! With more handsets on the market now and towers being brought online every week, users will start to see some real-world advantages of 5G. Up until now, it has been nice to show your friends an Internet speed test and hear the excitement when big numbers come up but expect to see more applications of the speed and low-latency that 5G offers.

With less car travel occurring around the world, you might be mistaken in thinking that the development of electric cars (EVs) may slow. I believe the contrary. Progressive governments in some countries have shown that users respond to a slight nudge in the direction of EVs. This country has some catching up to do. Our tax system allows a 2.6 tonne, 5.7 litre V8 ute consuming over 12 litres of fuel per 100 kilometres and costing in the vicinity of $200,000 to be exempt from luxury car tax (LCT) but a $78,000 EV is charged LCT. Despite our laggard politicians, enough governments around the world see the (electric) light at the end of the tunnel and different manufacturers are developing new technologies at an impressive rate. It still won't be the year for the flying car but the future is electric and we will see a dramatic increase in EV uptake across the world.

Notable mentions are wearables; robotics; more efficient solar PV (including black silicon cell technologies) and increased surveillance usage – by both governments and individuals.

Tell me your number one technology prediction for 2021 at ask@techtalk.digital.
TECH TALK PREDICTIONS: 2 JAN 2021

Opinion Articles

As a result of the technology columns I write, I am often granted permission to write random opinion pieces. Some are related to technology but sometimes I let my mind wander.

These articles explore the worlds of; EV, climate change, sport, laws and the way the world works...and, of course, technology! Now, I am not an expert in all of these fields, but these are simply my opinions towards hot topics in our society.

Enjoy these stories and let me know if you have similar opinions or a different perspective!

Busting the EV Myths

Time for me to do my best impression of Adam Savage. I want to bust a few myths today. There seems to be so much discussion around the technology associated with electric cars at the moment – and I struggle to find data and facts in amongst the spin.

I am not formally qualified to add research data to this debate but I feel I have enough experience with ground truths to add some common-sense and critical thinking. From an early age the concept was always instilled in me to use utilise research and data rather than fanciful opinion when it comes to presenting a point of view but I also have a long history with electric vehicles.

I bought my first hybrid fifteen years ago and I have used a total of seven fully electric or electric hybrid vehicles as my main vehicle in the ensuing period. On 18 August 2005, a former Fairfax editor, Linton Besser, wrote a story on my push to have Dubbo City Council use hybrid vehicles and I was the first Mayor in the nation to use an electric vehicle as the main Mayoral vehicle. I currently drive a Tesla which I purchased in March last year and it now has over 40,000km on the odometer.

So with that background, we should explore some of the wild statements being made by people either displaying ignorance or with vested interests.

Myth 1: Charge time.

I have heard anywhere from 5 minutes to several days quoted. There are a variety of types of chargers and companies providing charging infrastructure. Tesla opened a Supercharger station this week in Dubbo and they are the best known with 12,888 Superchargers across the world. Tesla chargers will only charge Tesla cars but there are a variety of universal charger networks that will charge all electric vehicles (including Teslas) with appropriate adapters.

You have different connectors – such as CHAdeMO and J1772 and Type 2 and Tesla – and different suppliers of the infrastructure – such as Tesla and NRMA (one to be opened in Dubbo shortly) and Chargefox and more. Some are free to charge and some charge a fee that is comparable to your home electricity costs.

The all-important question though – how fast?

Just like your mobile phone, the charging rate slows as it nears full capacity but my car will charge at a Supercharger at 600km/hour. I pay nothing to charge at a Supercharger. At home I charge at 80km/hour and I use solar panels to charge BUT if I was paying for electricity, it would cost about $15 to 'fill up.' New V3 Superchargers are being rolled out by Tesla which will charge at 1,500km/hour.

Myth 2: Range.

One of the issues with some of the early electric vehicles was range. An earlier electric car I drove had a range of 170km. Perfect for my job in attending to events in Dubbo but not great if I needed to drive to Sydney daily.

A reduction in battery price and increased efficiency means that the range of electric vehicles is constantly improving.

My current car has a range of 632km on a single charge – even further if I drive slowly. The average Australian drives 15,530km per year or just under 300km per week. Many reasonable priced new electric cars are coming out with a range of over 270km per charge. For most people they can then charge them once or twice a week. There are also many examples of people that have driven around Australia in electric vehicles. Charging stations are typically 200km apart.

On a long trip you can drive for 2 hours and have a 20-minute break while topping up. There are currently about 6,400 petrol stations across this nation but only about 1,000 electric charging stations. It is early days and that ratio will change. I think you will start to see some of the petrol stations adding electric charging stations as it is a logical step for them. This lack of charging infrastructure is a current weakness but one that will be addressed over time.

Myth 3: Cost.

Many people point to a Tesla Model S – with a starting price of $120K – as the prohibitive factor in purchasing an electric vehicle. There will be at least eight all electric cars available in Australia by the end of the year and some will start below $50,000.

That may still seem expensive, but think of the total cost of ownership.

Over a five-year period, the cost of owning a vehicle is the purchase price minus the resale price plus the ongoing fuel and maintenance and insurance/registration costs. At the average distance travelled by a vehicle with average fuel economy and average petrol prices with CPI built in and adding in regular maintenance, an internal combustion engine (ICE) car would cost almost $16,000 in ongoing costs excluding insurance and registration.

There is more to wear out on an ICE car as well so the resale value of a car with almost 80,000km on the odometer would be less. An electric car, assuming similar insurance and registration costs, would reduce the $16,000 running costs dramatically and would lose fewer dollars in resale value. Suddenly the $50,000 electric vehicle stacks up quite well against a $30,000 ICE vehicle.

Myth 4: Utes and performance.

I was once a hot-headed young twenty-something with my V8 253 Holden HZ Ute with a modified exhaust that probably damaged my hearing when I put my foot down. The Toyota Hilux Ute has been quoted as the most popular vehicle in Australia and the accusation is that if we go to electric vehicles it will destroy modes of transport for tradies and for people who want to have weekend fun!

How ridiculous.

My old ute could do 0-100 in 9.1 seconds. The Toyota Hilux struggles to get below 10 seconds. One of my previous electric cars was a Nissan Leaf which was purchased for under $30,000 and it had 0-100 time of less than 8 seconds. The

Tesla Model S can do 0-100 in 2.6 seconds.

There are people in our nation right now who have done manual conversions of their Toyota Hilux from ICE to electric. The performance is fantastic and the range is only limited by the amount of battery power provided. Electric cars can have towbars and can be just as much fun as an ICE.

This is probably one of the silliest arguments put forward so far.

Myth 5: Percentage of sales.

I do some talks for different groups wearing the hat of a futurist. In those talks I predict that, regardless of government intervention, 20 per cent of new car sales across the world will be electric cars by 2025. In the current debate, there has been a target of 50 per cent of all new car sales to be electric by 2030. Both figures are eminently achievable. Across the world, new car sales for 2018 show fascinating results.

Norway already sits at 49.1 percent. Iceland is at 19 per cent. Sweden and the Netherlands are at 8.2 and 6.5 per cent respectively.

While the US is only at 2.1 per cent, California has specific incentives to reduce air pollution and, with a population of 40 million, sits at 7.8 per cent of new car sales. Australia, through a distinct lack of leadership, sits at 0.2 per cent. This is one of the lowest in the world.

That will soon start to change.

Myth 6: The electrical network.

The argument is that if everyone changed from ICE to electric cars, the network couldn't handle the load.

We need to look at the numbers.

There are 1.1 million new cars sold in Australia each year. If 20 percent were electric, that would be 220,000 in one year. Electric cars use about 16kWh per 100km. At the average distance driven, 220,000 cars would add up to 3.4 billion kilometres driven. That is a total of 512GWh of electricity.

In comparison to Australia's total electricity consumption, that is in the vicinity of 0.22 per cent. Put another way, the 33 wind turbines at Bodangora produce 0.19 per cent of the electricity used in Australia.

We have six years before 2025 when we will hit 20 per cent of new car sales and we will only need to add 0.22 per cent to total production of electricity of each year we hit our 20 per cent.

It will be some time before all 19.2 million cars on our roads are electric but even if every single car magically transformed to electric tomorrow, total electricity consumption would add less than 20 per cent to our current usage.

Don't forget that oil refineries use power as well so if all oil refineries stopped producing oil tomorrow, there would be more power available, which leads to my next myth.

Myth 7: Power consumption of refineries.

It takes 7kWh to refine 5 litres of petrol for an ICE to travel 47km. An electric car can travel 44km on 7kWh. Stop refining petrol and the grid will have all the power you need. I have researched the data on this from many angles and, although the 7kWh for the refining of 5 litres of petrol is about right, the 7kWh does not all come from the grid. There is power used by a refinery that comes from the crude itself and transportation that is not from the grid and losses in energy efficiency so a more accurate figure might be 1kWh of actual electricity provided by the grid for 5 litres. A more accurate comparison might be the energy saved by not refining oil might translate to 6km of travel not 44km of travel. A saving nonetheless but not a complete replacement.

These are the most common myths that have been thrown around. Look on some EV forums or talk to some EV owners and find out the facts to satisfy yourself that an EV world is the world we will end up with – not necessarily through government intervention but by simple market demand. PUBLISH DATE: 12 APR 2019

ODI DATA

After 4192 ODIs does data support what we know to be true. One Day International (ODI) Number 4192 was played on 14 July 2019. This was a milestone match. It was a World Cup final. After ending in a tie, it was the first ever ODI to be decided by a Super Over. And then it wasn't really decided by a Super Over with the Super Over a tie. It then also became the first ODI to be decided by a boundary countback – and after the uproar from fans and players around the world, I dare say it will be the only ODI ever decided by a boundary countback.

It was also about time England had a victory.

This was the twelfth edition of the World Cup and, after having played in the first ever ODI back on 5 January 1971, England had never won the World Cup. Australia played England in that first ever ODI and they have racked up five World Cup victories so England have some catching up to do.

With all of that background, I thought it was time for me to have a look at some statistics from those 4192 matches and see if there is anything interesting in there.

When ODIs started, players representing their country in Test matches were handed the responsibility of playing for their countries in ODIs. It made sense. There was a bat, a ball and a cricket pitch. As teams quickly started to learn, five days of playing with no maximum number of overs to face is actually a little different to a maximum number of overs per team with the match concluded in the one day. In 1971, New Zealand's average run rate in Test matches was 1.96. Australia and England, the combatants in the first ever ODI, scored in Test matches in that year at 2.32 and 2.54 respectively. The West Indies were virtually swashbuckling in 1971 scoring their Test runs at the rate of 2.73 runs per over (RPO). Note that all scoring rates are adjusted to equate to a 6-ball over with some countries employing 8-ball overs at the time.

The players had calculated that scoring faster was essential and the average run rate for the match in the first ever ODI was 3.84. Test opening batsmen went into bat in the ODI and took the shine off the new ball and played a slightly more attacking version of their normal cricket. In the first ten years of ODIs with more than two hundred innings played, only three times did a team manage to score at a run a ball or better.

The first thing that people talk about in the modern game is the faster scoring rate.

With boundaries now smaller and a six requiring the clearance of a ten-centimetre advertising cushion set in three metres from the one-metre-high fence that used to be the boundary, it makes sense that more boundaries are going to be scored – particularly sixes. Add in the fact that bats have progressed in their technology and pitches are prepared for runs not wickets and it seems obvious that run rates are going to increase. Then there is the Sri Lanka factor. I remember watching ODI matches in the eighties. After fifteen overs, it seemed that a score of sixty with hopefully the loss of no more than two wickets was a team on the move. Remembering that a score of two hundred was typically seen as a winning score. Then came Sri Lanka – and in particular Sanath Jayasuriya. The Sri Lankans took the approach that the first fifteen overs was an easy time to score runs with a harder ball and the fielding side restricted in the placement of fielders so they attacked from the first ball. In what was seen as a major upset, the 1996 World Cup final was played between Australia – already a runner-up and a winner in previous World Cups and perennial easy beats, Sri Lanka. The combined average of the opening pair for Australia was 36.97 compared to Sri Lanka's inferior 29.45 but Sri Lanka was more intent on attack rather than preservation. The combined strike rates told the picture of the progression in the game. Taylor and M. Waugh for Australia struck at a combined average of 70.8 whereas the Sri Lanka opening pair of Jayasuriya and Kaluwitharana managed to score 87.9 runs for every 100 balls. This started a revolution that is still being enhanced today with the endless pursuit of rapid runs to start an innings.

But what do the overall averages tell us. From 1971 through to the World Cup final, the average runs per over for all innings across all matches has seen a steady increase. Not many matches were played in the first few years but the average per calendar year for the first few years was 3.84 for 1971 then 4.04; 3.51; 4.65 and 3.95 for the next four years. It was not until 1980 that the average stayed above 4 RPO for all subsequent years. My recollection is that a score of 200 at a run rate of 4 was a typical target for any team. Keep in mind that this is when the West Indies were running rampant. During the early eighties, they controlled the world of ODIs like no team before or since but still the average run rate across the world only reached a peak of 4.67 in 1982. In the seven-year period from 1980 to 1986, the Windies had a win percentage of 79.4 per cent and a five-year win percentage of 81.2 per cent. The only other countries to come close to that are Australia in 2005 with a winning 78.3 per cent over seven years and South Africa in 2001 with 75.2 per cent. Current World Cup winners, England, only sit at 59.9 per cent over the past seven years and India in 2011, after having won the World Cup in that year, only sat at 60.4 per cent. The Windies had some serious firepower during this time – Viv Richards; Clive Lloyd; Gordon Greenidge; Desmond Haynes; Richie Richardson...they just kept rolling out batsmen made for ODIs. During this time of Windies dominance, Richards averaged 50.7 and struck at 90.7 – these are impressive numbers even by modern standards.

With the average RPO in 1980 at 4.13, the average for all matches per calendar year bounced around between 4 and 5 until 2005 when it finally made it over 5 with an average for the year of 5.1. The calendar year average hovered around the 5 mark – sometimes just below and sometimes just above until 2011 when the average for the year was 5.04 and it hasn't dropped below 5 since. The highest average for the year so far has been 2019 (partial) with 5.53. On average, the run rate has increased by 0.034 runs per over per year in the life of ODIs – but as with any average, there are some dramatic highs and lows. If we look at individual innings along that steady growth, there are some extreme outliers.

On the plus side, 2007 saw an innings at a run rate of 15.83. In 2015; 2014 and 2003 there were other individual innings with a run rate above 10. Similarly on the low side, in 1979 a completed innings showed a run rate of 1.11 then in 1992 we saw a run rate of 1.68 and in 2009 we witnessed the third lowest completed run rate of 1.77. Did this mean that suddenly run rates were on their way down? It was only two years earlier that we saw the highest individual run rate but, despite this low run rate in one innings in 2009, the annual average for 2009 was 5.12 compared to the annual average for 2007 of only 5.04. In any information such as this, individual information does not tell the story as well as the overall average or trend. Some people see a record low temperature and immediately declare that Climate Change is all bunkum or a one-off scorching hot day and declare the proof that is required for Global Warming but the worrying trend with Climate Change is the average increase in the temperature of the earth over more than a century of measurements.

With so many matches played, is there a clear indicator of who will win a match based on items other than skill and commitment of the players.

Win the toss? If your captain wins the toss of the coin – which should be a piece of pure luck with a fifty-fifty outcome, then your chance of winning the game is...50.1 per cent. Pretty much the same as the luck of the way the coin falls.

What about batting first or second? After the advent of T20 cricket, the mantra from many teams is that they want to bat second as they feel confident in chasing down any total. What does the data say? In all matches played, if you bat first, your win percentage is 48.6 per cent so the chasing hypothesis has some credibility but if we break it up into daylight matches and day/night fixtures, we see some different data.

If the match is played entirely in the day, batting first drops your team to a 46.4 per cent winning result so there is a definite advantage in batting second – in the day time. In day/night matches, the advantage flips. Batting first when the lights will be needed suddenly increases your winning chances to 52.3 per cent.

I started looking for other components outside of the skill of your players and there is one component that stuck out as an obvious one – but the data supports what most people know. The home ground advantage gives you a 58.2 per cent winning chance.

Now to combine a few of those aspects. If you play at home, you win the toss, it is a day/night match and you bat first, you have a 64.6 per cent winning chance.

Then I thought I should add some skill to the equation.

The average RPO across all completed ODI innings from 1971 through to the World Cup final has been 4.84. That equates to a score of 242 in 50 overs. We need to do just one run better than average for my first calculation. Based on the previous factors, if you play at home, you win the toss, the match is a day/night match and you bat first PLUS you score more than 242 runs, your winning percentage jumps to 78.0 per cent. Take the same equation but this time increase your score to more than 280 runs and your percentage just jumped to 84.1 per cent.

Go figure – it seems that skill and performance dramatically outweigh factors somewhat outside your control – such as the way a coin falls. Maybe that is a little like life – many people blame their outcomes on luck or external influences when the reality is you just need to score the runs yourself!

Tell me if there are other factors you think I should have considered.

Mathew Dickerson is not a retired Test cricketer or someone who has played sport for his country. Describing Mathew as an average club cricketer is an insult to the standard of club cricket across this nation. Mathew has played and watched and been interested in cricket his entire life – and has an opinion. PUBLISH DATE: 21 JUL 2019

WHICH LAWS DO WE IGNORE?

We have a funny legal system in Australia. Most of us have broken some laws but we don't look over our shoulder every minute of the day worrying about the long arm of the law. Some of us are even breaking laws on a daily basis.

If you live in Victoria and you have ever changed a light bulb, unless you are a licensed electrician, you have just broken the law.

Staying in Victoria, if you fly a kite or play a game in a public place 'to the annoyance of another person' than you have committed an offence. I wonder if I can get my kids arrested at the park when they annoy me? Walked on the right-hand side of the footpath?

Annoying? Yes! Illegal? Apparently yes as well!

My Mum broke the law more times than I could count. She would always leave the keys in her car because she trusted people. Unfortunately, that is illegal along with leaving windows wound down more than five centimetres when the car is unattended.

Paying for your food or coffee at the drive through window using your smartphone is illegal and can cost you points on your license. The Police could have a field day parked at any fast-food outlet in Australia!

When your kids first learn to say 'Look Mum, no hands' as they ride their bikes with no hands, they have also committed their first illegal act.

The list goes on. Don't even get me started on halfbacks breaking the rules of the NRL when they feed a scrum – and the referee is standing right beside them!

Luckily we have humans - not robots - working in the police departments around the country and they can use their discretion when they see a crime committed that seems a little archaic or silly.

But should that be the case? If I was booked after a police radar caught me driving at 101 kilometres per hour, it would be a story at the pub about what a terrible person the officer was. I was just over! But I was still over. On a lucky day, I might get away with ten or even fifteen over the limit but I could technically still be booked for driving at 101.

This is the current situation we find ourselves in with abortion in NSW. The current debate raging in our Parliament was started by an Independent politician who saw that the law and the practice were vastly different. Alex Greenwich introduced the Reproductive Health Care Reform Bill 2019 to Parliament to provide clarity around this emotional issue.

We have heard the discussions that have been occurring. Abortion is illegal in NSW. Yet there are approximately 25,000 abortions performed every single year in this State. How is that possible, I hear you ask. In this century there has only been one case of a person in NSW being convicted for a crime related to abortion. If the 'crime' is committed 25,000 times a year, surely we would see more arrests?

Well, despite the fact that an unlawful abortion has been a criminal offence since 1900, you can have a lawful abortion if a doctor believes that your physical or mental health is in serious danger by continuing the pregnancy. That serious danger can be any economic, social or medical reason.

So, despite the fact that abortion is listed under the Crimes Act, the reality is that every abortion performed in NSW would have a doctor saying that it was necessary to preserve the woman from serious danger.

The everyday practice seems to be somewhat at odds with the intent of the law –

hence the sensible approach to make the law clear for everyone.

I am not saying which side of the debate I sit on but there have been some interesting discussions around the exact timing for an abortion.

I will start with a basic assumption. I will assume that everyone would agree that killing a baby one millisecond after birth would be murder and one millisecond before conception I will assume we all agree that no human exists.

A large part of the debate then seems to have focused on when, during that nine months, it is acceptable to have an abortion.

If I go back just a few years, Aristotle postulated that an embryo gained a human soul at forty days for males and ninety days for females. Before that point, Aristotle believed the embryo did not have a human soul so presumably abortion was acceptable before those points and no, I also don't understand why a female took so much longer to gain a human soul.

Many of the discussions focus on the first trimester of the pregnancy. If we look at statistics from South Australia, ninety-one per cent of all abortions happened in the first fourteen weeks. Another train of thought discusses abortions up to twenty-two weeks. Again using South Australia as an example, just over two per cent of abortions occurred after twenty weeks. Other arguments talk about the time when a baby has a heartbeat being the cut-off – but this is incredibly early in the pregnancy – usually around eight weeks. I imagine that this particular aspect of abortion will consume many of the discussions. After conception and before birth gives a range of possibilities and there is no obvious answer for any particular point in time.

When can we say for certain that a human exists along that timeline?

Those arguing against abortion say that a human exists at the moment of conception and the law needs to stand up for the unborn child who has no voice. But what if that birth brings a child into a world of abject poverty or into a broken marriage or the pregnancy was as a result of a rape? I am not trying to answer these questions but merely pose them as thinking points. When debate and arguments become very emotional and heated in both directions, it is vital that those putting their views forward can also see there are other points of view. Many in fact.

I have even heard it said that the legalisation of abortion will lead to a more promiscuous society with no care for pregnancies because abortions can easily be obtained but with our myriad of options for contraception I can't see legalisation suddenly spurring us into a rampant sex society.

You will no doubt have your view on the entire debate but I congratulate Independent Alex Greenwich for bringing this debate forward and for the sake of consistency, allow the laws of the State to be determined and then enforced – whatever the final decision is.

Mathew Dickerson is not a lawyer but has attended the stage production of Legally Blond! PUBLISH DATE: 6 AUG 2019

Electronic voting? I would vote for that

When the fictional character Tom Dobbs (played by Robin Williams) was elected as President of the United States in the 2006 movie, Man of the Year, it played on the fears of people across the world when the subject of electronic voting is raised. In the movie, the Delacroy voting company won the contract to provide voting machines for the Presidential election but a software glitch delivered the winner based on spelling rather than votes.

> Across the world, only fourteen countries have dabbled in electronic voting and, of those, it is only widespread in four.

With the greatest respect to the various commentators and election analysts, one of the most attractive components of electronic voting is the instantaneous and incredibly accurate method of delivering a result. In recent Federal and State elections in this country, after watching hours of analysis by a variety of experts and witnessing numbers dribbling through into tally rooms, we all went to bed on the Saturday night not entirely sure who was going to be in charge. A further attraction is the dream that we won't waste tonnes of paper for candidates to print beautiful glossy brochures to hand out to voters as they make their way through the scrum of volunteers.

The positives are obvious for all to see so why is it that we are reluctant to go that way. Australia in particular, with residents known for being early adopters of technology, logically would embrace electronic voting.

The fear of security breaches seems to overcome the attraction of the advantages. There may be other reasons our governments don't head down the path with electronic voting but security is the main issue put forward.

The US has long had a variety of solutions to this problem and, via DARPA (Defense Advanced Research Projects Agency) they have recently awarded a $10 million project to an organisation to create an open-source voting machine that could prevent hackers from tampering with votes.

The first real test occurred at DEF CON 2019. DEF CON is the world's largest hacker convention with over thirty thousand people in attendance.

At the Voting Machine Hacking Village setup in DEF CON since 2017, hackers typically take a few minutes to expose vulnerabilities in voting machines. At the 2019 event, the latest hardware from DARPA was on display at DEF CON and, ignoring a few minor setup issues, no hacker was successful in breaching the voting machine.

With Trump's presidency still plagued by rumours of Russian interference, DARPA is keen to produce a system that will allow lawmakers to pass an election security bill in time for the 2020 presidential election.

I would also suggest there is a large potential return on investment if they manage to produce a secure system. Once security was removed as an issue, governments around the world would be interested in licensing the technology to use in their elections.

And maybe there is something there for a progressive Australian government (oxymoron of course) to look at.

It is common knowledge that the CSIRO has generated over half a billion dollars for Australians from its Wi-Fi patent. I would propose that we could task the CSIRO with creating a secure voting methodology, patent it and use it in elections in Australia. Licensing that technology to other governments around the world to use in their elections would surely generate significant revenue for our government?

Perhaps I am over-thinking the first step though. Much of the focus for electronic voting is on securing the process for remote voting. I would propose a staged process to introduce electronic voting.

Stage 1.

We continue to use voting booths and have people physically visit an election booth on election day with one significant twist. Instead of printing reams of paper and asking voters to write on those pieces of paper, I propose that we introduce electronic voting terminals into all polling booths. When you enter a polling booth in the normal way, you need to produce ID and you are electronically recorded as having voted. This immediately is more secure than the current method. Currently you don't need to show ID and you could enter every different polling booth during election day and your name would be ticked off in a book. After the event you may be contacted and told you voted more than once but with no ID check, where is the proof that you actually did vote more than once?

Once your name was marked off, you go to an electronic terminal to vote for the candidates. Being electronic, you would receive a warning if your vote was not completed correctly. If you are going to make the disappointing deliberate choice of voting informally, you can still choose to do so but accidental informal votes would be a thing of the past.

You finalise your vote on screen and 'submit' the vote. Once it is entered, two pieces of paper print out from your terminal. These two identical printouts have two pieces of information on them. Firstly, it shows how you voted. Secondly, it creates a unique random number on the voting slip. As you leave the polling booth, you drop one slip into the ballot box the same as you would do now and you put the other piece safely in your handbag or wallet.

At the end of the election day, all of the votes are already recorded and it would take mere seconds for all votes and preferences to be counted. The results would be known immediately after the election.

If there was any suspicion of a breach of the computer system, the physical pieces of paper would be on hand to allow a physical count to take place as well. Furthermore, a Web site would publish ALL votes on a Web site with the unique random number identifying how each vote was cast.

There is nothing to identify an individual by name with those details but any voter who was concerned about the legitimacy of the process could look up the information on the Web site and check their unique random number against the piece of paper they took with them from the voting booth. This would satisfy each individual that their vote was accounted for correctly.

At this stage the main advantage gained is that the votes are entered quickly with less chance of fraud and the count is instant. Keep in mind that with the current physical voting scenario, all of the information is entered (by humans) into computers and then they are instructed to perform the count.

Stage 2.

Once the voting public was familiar and comfortable with this method, the process could be extended to remote electronic voting. Using the same processes – ID check to access a portal then voting and printing a unique identifier etc. it would allow people to vote from the comfort of their homes – or from anywhere in the world, in fact.

We are now at the stage where we are comfortable with our entire life savings being accessible from our smartphone and our lives seem to be contained within that little device in our hands – yet we don't seem to be able to vote electronically.

I am sure we can overcome this problem. It just needs the political will to make it happen.

Tell me if you would like to vote electronically at ask@techtalk.digital. Publish Date: 15 Aug 2019

FUTURE FOCUS

"Witchcraft to the ignorant...simple science to the learned." And with that simple statement, science fiction writer, Leigh Brackett, summed up much of the technological environment we see today. The more famous version of that sentiment followed on over thirty years later when, in 1973, Arthur C. Clarke's third law stated that any sufficiently advanced technology is indistinguishable from magic.

Can you imagine taking some of our modern conveniences back only a few decades and showing people of that time. You may very easily have been burned at the stake!

TVs with wireless remote controls only started appearing in the late seventies. Imagine magically making a TV change channel without touching it. Along a similar vein, remote controls for cars didn't start appearing commonly until the nineties.

The first mobile phone call was made in 1973 and commercial TV itself only appeared in this country in 1956. You don't have to go back that far to appear magical by making a glass bulb turn on with the flick of a switch with Tasmania being the first state with a transmitted supply just over one hundred years ago.

There are so many items we take for granted and use every day that are only relatively recent in our history: The Internet; GPS; Streaming, Smartphones...my young teenage daughter rang me last week when I was at 39,000 feet. Knowing that I would incur the wrath of my fellow passengers by taking the call, I assumed it was an emergency as she knew I was between countries. I still think taking a phone call while flying at 0.85 Mach and 12km above the earth is pretty cool.

When I asked about the emergency, there was nothing to report. She was bored and making a call to a plane was no big deal to her!

Without some simple understanding of the technology behind these everyday occurrences, they all appear 'magical'.

It is with that backdrop that I want to look forward. Given the speed of change in the technological world and how silly some technology looks with the advantage of time (I'm looking at *you,* 3D TV) making technology predictions is a dangerous game.

We have only had the Internet in this country for thirty years, but we are on the cusp of having the Internet on, well, everything. With the progression of low power transmissions, advanced batteries and better connectivity, we can connect so many useful items to the Internet. We are already seeing connected cars with the ability to remotely monitor location and speed of a vehicle.

Interesting when you give your teenagers the keys to the family transport but imagine the increased efficiency in an organisation when all fleet vehicles are being monitored in real-time.

Add in a dash of AI and the huge amount of data can be used to track the efficiency of drivers and delivery runs across a company. This is not uncommon already for regional businesses that rely on delivery vehicles to track the exact whereabouts of their entire fleet. Knowledge is power and this delivers a better experience for the clients and is safer for the drivers - this is while we still have drivers but I will leave autonomous cars alone for the moment.

On the other hand, autonomous tractors with some sprinkling of AI are already making a difference to agriculture. Sitting on a tractor for hours on end is boring and falling asleep and running into a fence can be costly. With no traffic to worry about and an open paddock, autonomous tractors are already working the farms in regional areas. The next time you drive on a country road and see a tractor out in the field, have a close look to see if you can spot a driver. When you factor in the ability to monitor the soil and weeds below the tractor and change behaviour accordingly, you start to see some dramatic improvement in efficiency of cropping.

This isn't limited to tractors. Water is a precious resource and fertiliser and chemicals are expensive. Pivot irrigators are now able to install IoT sensors at strategic locations around their irrigated area and the centre pivot used to irrigate an area can then automatically adjust water and additives to specifically what is required at each area. Yield increases of twenty per cent have been achieved with a reduction in costs. One client I know couldn't wait until the mobile reception was good enough at their farm and installed their own farm-wide Wi-Fi system to ensure internal communications were at a good enough level to implement this technology.

This same technology is being developed for broadacre farming although the measuring is more often performed manually due to the larger area being covered. Give it time though.

As IoT develops and communication links improve, permanently installed IoT devices across the ground will be as common as star pickets across a farm.

And don't think cattle and sheep farmers are being left out of this. As weather patterns change and farmers are trying to extract the most from precious few resources, measuring and monitoring is the new farming mantra. Entire livestock farm management software companies are developing solutions to constant monitoring. Measure what you treasure. Gone are the days of putting your livestock in a large paddock with a dam full of water and coming back when they have fattened up.

Critical measurements are being taken of stock and they are rotated through specific areas of farmland. Water troughs are monitored remotely to ensure a constant supply of water is available. The ultimate aim is to develop the technology far enough to remotely monitor individual animals from a remote location anywhere in the world. In Australia we have land but not a lot of water. If we can be at the forefront of this farming technology revolution, we are well placed to maximise the efficiency of the land despite our lack of water.

It is somewhat obvious but regional communities have dramatically lower population densities compared to metropolitan areas. That is a social strength but it has been a technology weakness. When additional communication links are being built – either wireless or fixed – it is logical that a telco will build infrastructure where population density is greater. Better return on their investment. Logically you would think this would continue with the advent of 5G – but maybe we will be rescued by the sheep!

With 75 million sheep in Australia (and 26 million cattle) if we end up with 5G monitoring devices attached to all of these animals, telecommunication companies may focus their attention on regional areas first!

I may be dreaming but there is logic in the argument.

Much of the world focuses IoT on smart home devices and freight deliveries and bridges and physical items but there is a strong argument that our increased food production should be at the forefront of our technological thinking.

In the same way that the CSIRO has led the world in areas such as Wi-Fi and plastic bank notes, I would be happy with $150 million going to the Australian organisation (rather than NASA) to develop world-leading technologies in IoT for the improvement of product yields.

For the supporters of the CSIRO, that decision would be indistinguishable from magic!

Tell me where I have missed the mark with regional IoT at ask@techtalk.digital.

PUBLISH DATE: 1 OCT 2019

I GIVE UP

These are three words that I very rarely utter – or even think. It doesn't matter if it is in sport or business or Monopoly. I can barely remember a time I have given up – but the cost at the moment is simply too great.

I am losing friends and spending way too long arguing with people each and every day about nonsensical ideas.

> It seems to me that conspiracy theories are growing faster than election day promises – with about as much truth in both.

I can't answer why seemingly rational and intelligent people seem to believe that the moon landing was faked or that 9/11 was a controlled demolition arranged by the White House. They describe in detail the damage that Chemtrails are doing to our fellow humans while explaining why so many erroneously believe the Holocaust was real. Of course, the US government research program, HAARP, controls the weather (can I dial up some rain for tomorrow morning please) and the CIA manufactured HIV. I am not sure if it is the Illuminati or Lizard People who are running the world this week but I am sure if I asked a flat-earther I would have the answer.

I am not sure if I am more annoyed about the content of these 'theories' or the incorrect use of the word theory. If supporters of these notions were being more accurate, they would be called Conspiracy Hypotheses to be true to the scientific method – but I digress.

As much as I can't answer why so many people believe these ideas, other people are conducting significant research into the reasons why so many people are believing ever so more fanciful conspiracy theories. I could find evidence of the term first being used as far back as 1870 but the term became much more common in the middle of the twentieth century and then the Internet dumped a container of fuel on the flickering embers. What some of the research has shown is that the less control people feel they have over their lives, the more likely they are to grab hold of the latest conspiracy theory. On the flip side, if governments are more open and give people a greater sense of control, people are less suspicious and less likely to believe in conspiracy theories. Furthermore, people with paranoid and Machiavellian traits are more likely to latch onto these false ideas.

Whilst annoying, I have been able to cope with the discussions over coffee with friends or comments on social media in relation to these variety of old chestnuts. As silly as these 'theories' are there is no real harm being done and it is good to exercise the critical thinking part of my brain. Some even create some laughter which always seems like a good thing.

What has finally pushed me to throw in the towel though is the sheer volume of disinformation into what the conspiracy theorists believe is the greatest scam in history.

Anthropogenic Climate Change.

If you believe the sideline experts (ignoring those annoying scientists who want to use facts and data and...well...science) then the list sounds more fanciful than my previous list of conspiracy theories.

Global warming was created by the Chinese to make US manufacturing non-competitive; renewable energy manufacturers are pushing an agenda to make money; scientists lie to gain more research grants; some unnamed nefarious force wants to exert massive new controls over the populace; developers are trying to sell real estate promising water views in one hundred years...the list goes on. At least with the fake moon landing a plausible reason of national pride during the Cold War was given. Climate Change sceptics struggle to come up with any reason why.

None of this has helped me though. While sipping on a latte surrounded by friends, as fast as I can provide some facts to neutralise one line of thought, another denialist statement is made. “The climate has changed before,” one friend states with absolute confidence. “Glaciers are growing and polar bear numbers are increasing,” rolls out next. “Nature emits more CO2 than humans – we only emit 3.6 per cent of all CO2,” which is technically correct, but doesn’t tell the full story. “Plants need CO2 to live so more is good,” seems like a logical argument. “Just by breathing now we are contributing to the CO2 in the air. We should stop breathing if we want to stop Climate Change,” is another favourite by the deniers. “CO2 is not a pollutant – it is natural,” closely follows the comment on breathing.

And then to finish with the coup de grâce. “CO2 only represents 0.04 per cent of our atmosphere so how can such a small amount have any impact?”

One by one I counter each statement with science and data and logic – the antithesis of the conspiracy theorist. My friend will inevitably tell me they still don’t ‘believe’ in Climate Change.

I dramatically pick up my coffee cup. I hold it a metre above the table and exclaim loudly with a look of sheer determination, “I refuse to believe in gravity” and let go of the cup. It hits the table and shatters into a dozen pieces. I tell my friend that science doesn’t really care what I believe or, for that matter, what he or she believes. Climate Change is not a belief system. Climate Change is a scientific fact. While beliefs will do nothing to change the consequences, actions from individuals who accept the reality and want to deliver a better world for our grandchildren will change the outcomes.

I am now labelled a zealot and my friends no longer want to have coffee with me – if for no other reason than I have been banned from many local coffee shops for smashing their cups!

So I give up.

No longer will I be breaking coffee cups and arguing with friends. I will sit quietly in the corner and accept our fate.

I do have one request of the world though.

I live in a regional area and I like to ride my pushbike along our river cycleway and on our mountain bike trails. When business or pleasure takes me to larger cities across our planet, I still like to ride a bike around the streets of the city. I have been noticing more and more lately that I am finding it a little harder to breathe and sometimes the air stings my eyes. I also notice it is often hard to see the beautiful bright blue sky I have become accustomed to back home.

Could everyone just do me a small favour and reduce our pollution? Maybe just start using more renewable power – such as wind and solar. It seems like they will be cheaper in the long run anyway. Cut back on those coal-fired power plants. Then move to electric vehicles a little quicker. When I ride my bike behind a car – in particular a diesel – it is really hard to breathe. Transform our transportation away from the internal combustion engine – and electric vehicles are so much fun to drive. And could we leave a few more plants in the ground? They look nice and seem to do a pretty good job of helping with pollution.

So if everyone could just do that for me – reduce our pollution – I think the world would be a much nicer place.

Actually, now that I think of it, if I hadn't already given up on arguing about Anthropogenic Climate Change, it would be at about this point I would have said something like this: "Reducing the amount of pollution would also reduce the amount of CO2 in the air and that seems like a good idea regardless of your stance on Climate Change. This seems like the core of the solution to the Climate Change problem so we have an ideal win-win situation."

But I have so I won't!

Mathew Dickerson is not a Climate Change Scientist. Mathew takes his car to a mechanic, visits his doctor when sick and uses an accountant to lodge his tax returns. He also has confidence in the scientific process. He is applying logic to a problem that is discussed in society every single day. PUBLISH DATE: 1 DEC 2019

Have we seen enough yet?

There are times, in life, when the evidence seems to be so overwhelming that the answer is obvious – yet, when it suits, humans have an incredible ability to obstinately refuse to acknowledge the obvious.

Confirmation bias plays a huge role in the way we see the world. We buy a new red car and suddenly we notice that the 'majority' of cars on the road are red. The percentage of red cars didn't actually change but we are more conscious of the red cars. Our bias is confirmed every time we see a red car but we ignore the white; silver and black ones despite the fact they make up 22; 20 and 19 per cent of the cars on the road and red only sits at 8 per cent.

Let me digress.

As a cricket fanatic I was incredibly disappointed that Andrew Symonds was chosen in the Australian Cricket Team for the 2003 World Cup. Symonds made his debut in 1999 and, after a spectacular start, his times in the middle had been a mixture of brilliance and frustration. In my mind, with all the expertise of a couch commentator, it was time to move him on.

The selectors saw it differently.

The first game of the World Cup was against Pakistan. As the current Champions, it should have been an easy game for Australia. I remember it well. I was at my normal weekly Rotary Club meeting. No smartphones back in those days so you couldn't have a sneaky look at your phone to see the score. We finished the meeting and walked out to the bar. Australia was 4 for 86 with Symonds just walking to the crease. I said to all within earshot that it would soon be '5 for' as Symonds shouldn't even be in South Africa. When Ricky Ponting – our last real hope - departed with the score on 126 it was looking like an easy win for Pakistan. By the end of the 50 overs, Symonds had scored his maiden ODI century and left the ground on 143 not out. Australia had made a great total of 310 and went on to win the match easily – along with the next ten matches to take them to a World Cup victory.

Symonds averaged 163 for the tournament at a strike rate of 91 and took 2 for 7 in the final to help wrap up the trophy. Despite only averaging 23.8 across his first 54 matches, he went on to average 45.1 across his next 144 matches at a strike rate of 92.

Friends of mine who knew my opinion asked me after the breakthrough century if I had seen enough yet to reconsider my opinion of Symonds? No – it was just one match. They asked again after the World Cup. No – it was just one tournament. Again, the question was asked after his second, third, fourth, fifth and sixth centuries. His five wickets in a match. His naming in the ICC World ODI XI – twice. His highest vote count for ODI Player of the Year in the Allan Border medal. His rise to Number 8 in ICC rankings. His naming by ESPN in the greatest ever Australian ODI team. No, I repeated over and over – blinded by my confirmation bias of the occasional low score or pride that possibly my initial impressions of Symonds were wrong. When he managed a hat trick of ducks it definitely confirmed my negative bias – conveniently ignoring the unbeaten 175 runs he scored in the two previous innings at a strike rate of 117!

In hindsight it appears to me that the experts - otherwise known as selectors - seemed to know exactly what they were doing and, as much as armchair experts might disagree, we should have some faith in their expertise.

Back to my point.

We are having record days of heat in Australia at the moment. 16 December 2019 was the hottest ever day on record averaged across all of Australia which beat the previous record from 2013. That record lasted precisely one day with 17 December even hotter including the maximum temperature recorded at any single location with 47.7°C logged at Birdsville.

Big deal I hear many people say. We have had hot days before. It was 112 years ago when Dorothea Mackellar wrote her poem about a sunburnt country so a hot day is nothing new.

2019 was officially the hottest year on record in Australia with the temperature 1.52°C above the long-term average and globally the six hottest years on record have occurred in the last six years and the twenty hottest years on record have occurred in the last twenty-two years.

Yeh sure – that might look bad at first but are you sure about that data? Temperatures have only been recorded for approximately 140 years. Surely there were some hotter years prior to 1880 but they were simply not recorded and human history goes back a lot further than 140 years.

We currently are experiencing some of the worst bushfires ever with a record 6 million hectares of bushland destroyed this season so far – and we haven't even reached February which is typically the worst month for bushfires.

I hear the counter argument. Whilst unfortunate, bushfires have long been a part of the Australian landscape. Australia is a tough country and who can forget the Black Saturday bushfires on 7-8 February 2009 where we lost 173 people and 2,000 homes or the Ash Wednesday fires on 16-18 February 1983 where 75 people died and we lost 1,900 homes?

While Australia is burning, Thailand is flooding. The New Year saw Jakarta experience 377mm of rain in a single day – the highest level since records began. 21 people have died and 30,000 have been displaced.

Well Jakarta has had heavy rainfall before – in 2007 and 2015 for example. The country's Meteorology, Climatology and Geophysics Agency said this is "not ordinary rain" but surely it is just one of those freakish acts of nature.

We could use some of that rain in Australia. The national average rainfall total for 2019 was 277.6mm. Is that low, I hear you ask? The previous record low was 314.5mm so this year was 11.7 per cent lower than the previous all-time record.

Ha! This is just way too easy for the deniers. Not even a challenge. "This is simple high-school science. When it is hotter there is more evaporation. If the planet is warming then more evaporation will lead to more rainfall. QED!"

2019 saw another record high for the daily average CO2 level with 415.70ppm recorded in May – but this will be a record held for only a short period with CO2 rising at almost 3ppm annually. This is up from 291ppm in 1880.

That CO2 thing I hear some say. That is completely covered. That is just a natural cycle of ebbs and flows of CO2 in the atmosphere. Nothing to see here. We have lots of examples in our history where the concentration of CO2 in the atmosphere varied by 100ppm. Sure, it took 100,000 years for that change to take effect but this one is just a faster cycle. Next!

I can keep going with many other data points but hopefully you see my point. Despite the overwhelming evidence to the contrary, if you don't want to believe the data, if you don't want to believe the facts, if you want to ignore the experts, you can always come up with some way to justify your position on anything. Whether it be red cars or Andrew Symonds, you can pull out selective data to make yourself feel comfortable with any position you take.

The problem with Climate Change though is that the earth doesn't care about your justification. It doesn't care what you think. It doesn't care about your 'belief'. The earth will follow scientific laws and respond to the inputs it receives. It has been doing just that with Climate Change since the Industrial Revolution and will continue to do that until...

Have we seen enough yet?

...until leaders across our nation, leaders across our world, finally say they have seen enough, seen enough of the indisputable data, to start to take action. Real action.

It took an unreasonable amount of evidence on Andrew Symonds before I finally admitted that I had been wrong about his contribution to Australian cricket. Sorry Andrew!

With the greatest respect to Andrew Symonds, the issue of Climate Change is of slightly more importance than my opinion on his contribution to Australian cricket. The evidence we have on Climate Change is more than overwhelming. It is irrefutable and cataclysmic if we ignore it yet that is exactly what we seem to be doing. Ignoring the mountain of evidence.

Surely it is time for our leaders to say that they have seen enough and take action. For the sake of our future, I hope so.

Have you seen enough yet?

Mathew Dickerson is not a Climate Change Scientist. Mathew takes his car to a mechanic, visits his doctor when sick and uses an accountant to lodge his tax returns. He also has confidence in the scientific process. He is applying logic to a problem that is discussed in society every single day. PUBLISH DATE: 3 JAN 2020

WHY DID WE BOTHER TO BAN CFCs?

Chlorofluorocarbons. CFCs. The word doesn't roll off the tongue - and it would be safe to say that before the eighties it was not a topic for common conversations at dinner parties. But in the eighties, we all suddenly became experts on CFCs and ozone holes.

Before we understand why, we need to go back to the beginning.

The 'father of the refrigerator' was Jacob Perkins who received a patent for a vapor-compression cycle using liquid ammonia in 1835. The liquid chemicals used in a refrigerator are called refrigerants and, to be effective, they need to satisfy several specific criteria including having a low boiling point.

Unfortunately, early refrigerators were appallingly risky because they used dangerous gases that were flammable or toxic or highly reactive – and they sometimes leaked.

Thomas Midgley set out to find a safer option and in 1930 US patent number 1,930,129 was granted for the manufacture of aliphatic fluoro compounds – or what we now refer to as chlorinated fluorocarbon. He demonstrated the effectiveness of his invention when he inhaled a large amount of the gas and then blew out a candle, showing it to be non-toxic and non-flammable. The substance was marketed commercially by Du Pont under the brand name Freon.

Freon and other CFCs soon largely replaced dangerous refrigerants and many other applications were found for the new inert gas with a low boiling point - such as propellants from items as diverse as aerosol cans to asthma inhalers and fire extinguishers.

The Society of Chemical Industry awarded Midgley the Perkin Medal in 1937 for his work with CFCs. And there ends a story of scientific discovery and progression in our society.

Except for a sting in the tail.

Fast forward forty years. James Lovelock used a sensitive instrument he had designed himself to test for the existence of CFCs on the west coast of Ireland. His hypothesis was that the haze in the air was caused by winds that blew pollution from the urban areas. If the haze was caused by pollution, it would contain larger concentrations of synthetic chemicals. CFCs would be part of this mix so he tested for the existence of CFCs. He detected CFCs in the air on hazy days and therefore confirmed its human-made origins.

Case closed.

But as any good scientist, he didn't just accept the initial results. He kept testing. And the results on clear days did not match his hypothesis. When the air was coming from the Atlantic with not an urban area in site, CFCs should be close to undetectable. They weren't. On pristine days, Lovelock was still able to detect CFCs. This started Lovelock along a different line of thinking. Were CFCs building up in the atmosphere everywhere?

Lovelock took his instrument on a sea voyage from England to Antarctica and took measurements along the way. Wherever he travelled, he found CFCs. Not a lot – but more than were expected. He found about 0.06 parts per million of CFC-11 in the atmosphere.

He presented his findings at a scientific meeting in 1972 and, while interesting, concluded that the level of CFCs constituted "no conceivable toxic hazard" despite the fact that practically all of the CFC-11 ever manufactured was still present in the atmosphere.

But the wonderful thing about science is that knowledge is cumulative. The dwarf can see further by standing on the shoulders of giants (nanos gigantum humeris insidentes) and two researchers, Sherwood Rowland and Mario Molina, wanted to further research the fate of CFCs in the atmosphere. When Midgley first set out to find a new refrigerant, he was looking for a compound that was inert so there should be no reaction with a CFC and the data from Lovelock basically confirmed that this was the case. That was in the lower troposphere. What happened when CFCs drifted up into the stratosphere?

CFCs would take some time to drift up as they are heavier than air. Air has an average molar mass of 28.97g/mol but CFC-11 comes in at 137.37g/mol. Despite this the atmosphere is a moving, living, dare I say it, breathing environment and gases are mixed together. Rowland and Molina calculated that CFC molecules would, over decades, wind up in the stratosphere where UV radiation would potentially split off chlorine atoms. Each chlorine atom would react immediately with an ozone molecule, they hypothesised, setting off a chain reaction that would destroy thousands of ozone molecules.

Their first paper that linked stratospheric CFCs and ozone depletion was published in 1974. It took twenty-one years but that paper would eventually earn them a shared Nobel Prize.

Rowland and Molina believed that the only solution was to ban the production of all CFCs immediately. Even if that occurred, ozone loss would go on for years. If CFC production continued, the ozone loss would be even greater.

Other scientists conducted further research testing the original hypothesis. Data and information varied with different testing methods and as scientists focused on different areas. Modelling showed varied timelines for different percentages of ozone being destroyed. One thing seemed certain – despite the variations it was clear that CFCs were damaging the environment.

Not surprisingly, there were some that weren't convinced that a ban on CFCs was the right action. Even less surprisingly, many were engaged in industries that relied on the production and use of CFCs. This was not a small industry. It was estimated that in 1974 industries relying on CFC production generated US$8 billion ($41.2 billion in current terms) and employed 200,000 people.

CFC manufacturers were trying to cast doubt on the scientific validity of the hypothesis in any way they could.

"It is just a hypothesis," they said.

"There is only flimsy evidence," they continued.

"There is no proof the molecules would ever reach the stratosphere," with the seemingly logical argument of the heavier than air weight of CFCs.

"Stratospheric ozone concentrations fluctuate naturally by geography and by season so any detected changes are just natural," again seemed quite logical as there are natural oscillations.

"CFCs should be regarded as innocent until proven guilty," was another line.

"812,522 tonnes of CFC compounds were produced in 1974 but the Earth's atmosphere is 6.3 billion times more massive. How could 0.000000016 per cent of the atmosphere have any negative impact," was another seemingly plausible argument put forward.

The industry even brought in their own expert to challenge the ozone depleting properties of CFCs with a sponsored speaking tour for Professor Richard Scorer, well-known for his research on pollution.

The fact that CFCs were seen to be such a highly useful class of chemicals and, particularly in refrigerators and air conditioners, no obvious alternatives existed, made it much harder for support of widespread bans. So the world continued to produce CFCs.

The scientific community realised, possibly for the first time, that our activities could impact Earth's environment on a planetary scale. It didn't matter where the CFCs were emitted, it was a global problem.

The US was the first to act. On 11 May 1977 the US government announced a timetable for phasing out of CFC in aerosols. Several countries followed suit. The ozone level was also going to be monitored more closely. World production of CFCs started to drop – but then started to grow again. Within ten years, the annual production had surpassed the 1974 level and was still rising. Molina and Rowland were advocating – still – for a total ban on CFC production from all sources. The impact of continued production would not be felt for many years, but they would be felt. Policymakers the world over were lagging with any significant response.

Now jump to 1984 and move to the cold climes of the Antarctic. More specifically the Halley Bay Station of the British Antarctic Survey. The levels of ozone above the Antarctic had been measured for decades yet in 1984 the scientists found that stratospheric ozone above the Antarctic had decreased dramatically. In September, in particular, when conditions were just right for CFCs to cause ozone depletion, the level of ozone was reduced by a massive forty per cent. The results were published in May 1985 specifically in relation to the ozone 'hole' above the Antarctic – which was later backed by NASA - and further studies showed that the average ozone level across the rest of the planet had decreased by possibly up to three per cent.

Without the continued protection from the sun's UV rays that stratospheric ozone offered, skin cancers would start to increase and many organisms would start to suffer genetic damage. If the ozone levels continue to be depleted, life on this planet would cease to be sustainable.

Finally...Finally governments put aside the short-term economic impact and looked at the future of the planet.

When there is the desire, the world can move swiftly to solve a problem. By 1987, the Montreal Protocol was agreed to by 56 countries. CFC production would be cut in half by 1999. Three years later this target was updated when a gathering in London saw 93 countries sign a treaty agreeing to phase out CFCs completely by 2000. Over the next two years more results indicated increasing ozone depletion hence the elimination date for CFCs was moved to 1996.

Today the spray paint and deodorant sitting on shelves is CFC-free. New refrigerators and air conditioners are using alternative refrigerants. The work of Rowland and Molina and countless other scientists was justified and acknowledged by the wider community. With the CFC-ban now in place, atmospheric levels of chlorine are beginning to decline and the ozone level is on the way to recovery. A full recovery is expected by 2070 – almost one hundred years after the impact of CFCs was first discovered. With hindsight it is easy to say we should have acted sooner but the world finally came to an agreement and we are on the way to recovery. The whole story does show how incredibly rapidly we can produce major changes to our atmosphere with seemingly small inputs and how long it can take nature to recover. Remember that ozone only makes up approximately 0.00006 per cent of our atmosphere and its peak concentration occurs at an altitude of 32km above the earth's surface.

Most importantly the world learned a few lessons and will obviously never repeat the same mistake...or...or is it already doing just that?

I am actually incredibly confused by the approach of governments across the world to ban CFCs. Not because it wasn't the right thing to do. Not because the evidence wasn't there. Not because it avoided a cataclysmic event that would end life on this planet as we currently know it.

No.

I am confused because we acted in a relatively short timeframe to resolve a world issue in relation to CFCs, but when a much larger potential threat looms across the planet in the form of climate change, we seem to be ignoring all the lessons from our near-miss with CFCs and continue blindly doing what we have done for over a century. What makes it even worse is that when the negative impact of CFCs was first discovered, one of the hardest issues to resolve was to find a suitable alternative.

With climate change we have the facts, we have the research, we have the data, we have the alternatives but we just don't seem to have the will. At least governments, in particular Australian governments, don't seem to have the will.

Why?

I could speculate that the US$1 trillion global coal industry and the US$2 trillion oil industry significantly dwarf the size of the CFC industry when it was threatened. I could further speculate that we have seen evidence of sophisticated misinformation campaigns by companies such as Glencore Coal with Project Caesar which used the C|T Group to run a vast, secret communications campaign over a two-year timeframe aimed at shifting public and government attitudes towards coal – at a cost of up to £7 million per year.

These reasons do seem somewhat feeble though compared to the threats that climate change is posing to our worldwide US$80 trillion economy and the lives and property that are already being lost. I just can't see why we aren't taking urgent and immediate action at every level.

I go back to my first question. Why did we bother to ban CFCs when all it seems to have done is delay the inevitable destruction that we will wreak on our only home?

Why bother fixing one puncture in our car if we are just going to continue to drive with a flat tyre the next time we get a puncture and destroy the rim.

Why?

Why?

I don't know why. Does anyone?

Mathew Dickerson is not a Climate Change Scientist. Mathew takes his car to a mechanic, visits his doctor when sick and uses an accountant to lodge his tax returns. He also has confidence in the scientific process. He is applying logic to a problem that is discussed in society every single day. PUBLISH DATE: 9 JAN 2020

CELTIC MEAGAN
AN EVIL GENIUS

One of my favourite intellectual TV programs when I was growing up was 'Hypotheticals' which aired on the ABC and starred Geoffrey Robertson. I loved the concept – put a hypothetical situation to a panel of experts and then explore the outcomes.

The topics involved ranged from historical dilemmas to political conundrums and more. One particular 'hypothetical' stuck in my mind. Robertson created a story where a Minister was reading confidential notes on the train and put them in his briefcase before briefly visiting the toilet. Robertson asked a journalist on the panel what he would do in this hypothetical situation. I was horrified when the journo said he would look in the briefcase and read the information while the politician was briefly away. This would never happen today of course – what politician uses the train?

> The beauty of posing any hypothetical situation – whether on the show or sitting around a coffee table – is that there are no actual consequences.

The premise and the actions are all imaginary but they can be incredibly interesting. I recently saw an episode of a Netflix show called Lucifer and there was a situation on the crime solving show that would make a perfect hypothetical situation. The criminal in one episode contacted a specialist in their field and gave them a tough choice. They could sacrifice a part of themselves that made them special or the mastermind would kill an innocent victim. For example the thoracic surgeon needed to destroy her hand; the actor needed to cut his face and the track athlete needed to amputate his leg. These body sacrifices would then save the life of an innocent University student.

What would you do given the same choice? It makes for an interesting conversation when you play with the moral dilemma with the added complication that you are being asked to trust a criminal to either follow through with the original threat or save a life if the requested action is taken.

While this type of hypothetical is one I would never like to be faced with, I want to play with another hypothetical today.

Picture a brilliant evil criminal – the likes of which you see thrown up in a variety of movies. I am thinking of someone with the combined brilliance of Lex Luthor; Professor James Moriarty; the Green Goblin and The Joker. For this hypothetical, we should call our nemesis Celtic Meaghan (obviously from Ireland with a name like that).

In typical fashion, Meaghan takes over the broadcasters of the world and sends her evil message. She has used her brilliance to create a system that will wreak havoc over the entire planet impacting every person across planet earth. She has been planning this for quite some time and has left no stone unturned in her quest to bring the world to its knees.

Her evil plan is designed to attack the world at a number of levels utilising no new celestial components but using what we already have in this world.

Unless Celtic Meaghan's demands are met, she has threatened to:

- Generate wild hurricanes;
- Start horrific bushfires;
- Spread infectious water-borne diseases;
- Create heatwaves;
- Destroy food supplies for developing nations; and
- Create cycles of dreadful floods followed by extreme droughts.

Meaghan, as we know, is evil. The 'evil genius' title tends to give that away. All of these actions she is going to undertake will take human lives and Meaghan makes no apologies for that. Meaghan has a very sick sense of humour. During the broadcast, she also informs the world that for some added 'fun' as she puts it, she will also:

- Destroy hundreds of animal species;
- Destroy UN World Heritage listed sites;
- Destroy coral reefs; and
- Destroy half the wine regions in the world.

As much as governments across the world say that they won't negotiate with terrorists, the threat from Celtic Meaghan is too great to ignore. The United Nations gathers to hear the demands of this crazy individual. They are ready to give Meaghan almost anything she requests to not inflict this pain on the world. Does she want all the Gold in the world? What about her own island to live on? Maybe a date with George Clooney. The world leaders would be prepared to remove their voice boxes to protect the world from this malevolent miscreant. What would you be prepared to sacrifice to stop the world from this death and destruction?

What will the sacrifice be in our hypothetical?

Celtic Meaghan has but one demand.

Stop polluting the world! Wait – what? That is it?

It turns out that, as well as being an evil genius, Meaghan is also an asthma sufferer and can't travel the world with all this pollution. She just wants a world with cleaner air. "Stop burning everything," she says. "All this oil and coal and gas that is being burned is destroying the beautiful fresh air that I love to breathe," she continued. Her message is quite simple – she wants a world free of pollution.

As it turns out, Meaghan is quite reasonable. Taking over broadcasting channels and threats to destroy the planet aside, what she is asking for actually seems like it would be OK for everyone.

As with any hypothetical, I have made up Celtic Meaghan and the scenario but don't we have a similar choice to make in our real world.

Climate Change threatens to do all that Celtic Meaghan was threatening – and more. If the solution to the problem was to sacrifice three billion people or listen to Justin Bieber music on constant loops, I can understand why many people in the world would be loath to solve the problem or deny a problem exists. When the solution is to reduce pollution, which sounds like it would be good regardless of the threat of climate change, I don't understand why many people and governments are resisting. Surely a policy to reduce pollution would be a policy universally embraced.

Please tell me what I am missing.

Mathew Dickerson is not a Climate Change Scientist. Mathew takes his car to a mechanic, visits his doctor when sick and uses an accountant to lodge his tax returns. He also has confidence in the scientific process. He is applying logic to a problem that is discussed in society every single day. Publish date: 10 Jan 2020

Y2K BUG? Y2 WORRY?

Take a moment to cast your mind back to the late nineties. We were obsessed with news about Bill Clinton and impeachment. We were horrified by the Columbine High School massacre. We listened to Elton John and Britney and a new generation became fans of the Star Wars franchise with Episode I 'The Phantom Menace' while the Euro changed the financial landscape across Europe.

In Australia, we debated endlessly about a lovely lady with a rather strange taste in hats being in charge of this nation. We were urged to vote Yes in that referendum by the Chair of the Australian Republican Movement who many of us suspected wanted to run the country.

And...Y2k bug fever gripped the world.

First some context. When computers the size of a room were being introduced in the fifties, storage was at a premium. The IBM 350, introduced in September 1956, had the world's first hard disk drive. At 971kg it could only store 3.75MB and would set you back US$330,000 (current equivalent).

In writing software, the cleverest programmers deduced that two bytes of data could be saved when writing the date.

For example, the date man landed on the moon could be written as 16071969 or 160769 with the latter more space efficient. To any human or computer, it was obvious they were both the same date and at up to US$90 per kilobyte (in current money), programmers were encouraged to be more efficient.

The Y2k problem was highlighted in the mid-eighties but the deadline was still some time away. Other problems seemed more important. By the mid-nineties, there was finally some acknowledgement that we should do something about the issue. Despite many computer scientists around the world running simulations and analysing millions of lines of code, the Y2k issue caused heated debates at coffee shops by people with minimal research on the issue.

Just for a moment ignore all the logical data from specialists and pretend that there were two possible outcomes. It was either a hoax or the greatest self-inflicted calamity in mankind's industrialised history. Then there were two possible actions. Throw resources at the problem to negate the possibility of this being the worst party to welcome the New Year ever – or ignore it.

That gives us four scenarios.

First outcome. Do nothing – and it was a hoax.

News reports: "Lucky we didn't waste any time worrying about that problem."

Second outcome. Do something – and it was a hoax.

News reports: "Much ado about nothing. We need a Royal Commission to find out who perpetrated this hoax."

Third outcome. Do nothing – and it was real.

News reports: There won't be any. A complete breakdown of modern society would ensure the least of our problems was the fact that we couldn't turn on TV and watch the news.

Fourth outcome. Do something – and it was real.

News reports: "Welcome to the New Year." You can hear the artist formerly known as Prince singing in the background: "So tonight I'm gonna party like it's nineteen ninety-nine."

Of the four available options, the world took the only sensible course of action available to it. If it had been a hoax – mind you the best orchestrated and well-organised hoax in the history of mankind by a factor of a bazillion – then we would have wasted some effort and money on tidying up some lines of code. If we did nothing and the world collapsed, with the knowledge we had, how could future generations ever forgive us for what we had allowed to happen knowing there was a solution.

Many Y2k deniers pointed to a functioning society after midnight and told us we all fell for a hoax. That ignores one little number. 500. That was the minimum estimation, in billions of US dollars converted to current terms, spent across the globe to ensure we were ready for the biggest party in the world.

And we still didn't get it quite right.

Despite US$500 billion, we still had date-related problems after 1 January 2000. Credit card failures; corrupted satellite data; school heating failures; nuclear reactor false alarms; rejection of food shipments; incorrect age-based screening tests for pregnant women and even a video rental late fee of over US$90K!

I can't help but think of the parallels between now and the late nineties.

Today we hear constant talk of impeachment of a US President and we witnessed yet another mass school shooting at Parkland, Florida. We are still listening to Elton John and Britney Spears. Supposedly the last Star Wars movie hit our screens with Episode IX 'The Rise of Skywalker' while the European financial landscape is being challenged with Brexit.

In Australia, Malcolm Turnbull did eventually get his chance to run the nation – just not as a republic.

And looming over everything we do – just as with the Y2k bug – is Climate Change.

As with the Y2k bug, I see four possible combination outcomes based on two variables. Climate Change might be real or it might be a hoax. Like the Y2k bug, the scientists and experts have provided the proof and simulations to show the harsh reality of the situation but just for the sake of the argument, pretend it is a possibility that Climate Change is a hoax.

We then have the option for the world to take action...or not.

Option 1. Do nothing. Climate Change is a hoax.

Result? None of the dire consequences of Climate Change come to fruition and we continue to live our lives as per normal. Sure, we will eventually run out of coal and oil but that is for another generation to solve.

Option 2. Do something. Climate Change is a hoax.

Result? We reduced pollution across the world while we spent unnecessary money on creating other ways to produce power and propel cars. Some wasted money but the world seems like a nicer place to live.

Option 3. Do nothing. Climate Change is real.

Result? We destroy the planet and kill most of the people living on it. Full stop.

Option 4. Do something. Climate Change is real.

Thank goodness. The world was on the brink but mankind saved themselves from themselves.

When you break down the options, is it really worth the risk? In just the same way as the world could not afford the risk with the Y2k bug, it simply can't afford to risk taking a chance on not acting on Climate Change.

The biggest difference between Climate Change and the Y2k bug is that there was a definite deadline that everyone identified with the Y2k bug. 1 January 2000.

With Climate Change, we may well have already gone past our irreversible deadline.

Mathew Dickerson is not a Climate Change Scientist. Mathew takes his car to a mechanic, visits his doctor when sick and uses an accountant to lodge his tax returns. He also has confidence in the scientific process. He is applying logic to a problem that is discussed in society every single day. Publish Date: 10 Jan 2020

Life is a Compromise

Every day we make decisions that involve some form of compromise. We have competing interests that we have to weigh up and then we make a decision. I like the idea of staying in bed all day and catching up on some movies but my kids need someone to get them organised for school.

Going to work today? Sounds lame. Surely taking a permanent holiday would be better – but I need to work to generate an income so I can take that holiday.

I know of an 85" 8K TV that would look great in my loungeroom but if I choose to spend $13K on a TV then I would need to choose to cut back on something else...like food! So, I buy a smaller TV...and I buy food.

We are constantly making decisions with how best to utilise our money and our time and, without realising it, we are constantly weighing up options with every decision we make.

When people see one of my electric vehicles (EVs) or talk to me about them, the common statement I hear is "I love the idea of an electric car, but I couldn't get one because..." and insert any one of a number of reasons. "I like a powerful car" or "I need more range" or "I don't know where to charge them" or "I live in a rental" or "I can't afford one" or...

Let me turn that argument on its head.

In an alternative universe, everyone drives an electric car. Someone turns up with a new car that has a flammable liquid stored under the boot. That fuel is burned in a combustion chamber to turn mostly into heat and noise and about thirty per cent of it in to forward motion. While sitting at a café talking to friends, one of them tells me about his new internal combustion engine vehicle (ICEV). I don't want to offend my friend, so I say, "I love the idea of an internal combustion engine car, but..." and I insert "I like a car with instant acceleration" or "I don't like standing around at a petrol pump wasting time filling up" or "I don't like all that noise of the engine" or "I can't afford that expensive fuel" or "I hate the pollution that the car is producing" or "I don't want to destroy the planet" or...

People make their decisions emotionally and justify them irrationally.

For our modern society, we become comfortable in what we are doing and we only want to make a change if it is going to make our life better today. The future can look after itself!

My household is now proudly free of internal combustion engines. It has taken me fifteen years to finally arrive at this point. In 2005 it started with a Toyota Prius. In 2010 I replaced the petrol lawnmower with an electric robotic lawnmower. In 2011 I installed 10kW of solar panels because I could see where I was headed and I wanted to make sure I could generate more electricity than all my electrical appliances would use. By 2012 I had my first car driven primarily by an electric motor – admittedly with a petrol backup engine to remove my range anxiety. My first fully electric car came in 2015 by which time I had replaced all my appliances – whipper snipper, chainsaw, etc – with electric versions. Having a household with one electric car and two hybrids for the last couple of years I have finally sold the hybrids and replaced them with EVs so my household now uses no petrol or diesel and instead houses three EVs.

Has that process involved compromises along the way? Of course. As mentioned, every decision involves some form of compromise. Were the compromises worth it? I would say absolutely. There is only one planet and with such poor government leadership in relation to the future of our planet I am convinced that the saving of our planet will not be led by government but it will be led from the ground up. By us.

Consider the following slightly oversimplified example. In Australia light vehicles generate approximately twelve percent of the 550 million tonnes of greenhouse gases we produce annually – or 66 million tonnes. We have 20 million cars in

Australia and we purchase around a million new cars each year. Forget government leading the way. Imagine if every individual that was going to buy a new car this year made a commitment to buy an EV. That would replace five per cent of our national car fleet and, if every purchaser of an EV committed to purchasing green power, that would immediately reduce our greenhouse gas emissions by 3.3 million tonnes. Do that for ten years and half of our ICEV fleet would be replaced and our production of greenhouse gases would be reduced by 33 million tonnes. Now that is only six per cent – but it is a good start. Under the Paris Climate Agreement Australia has an inadequate target of a 26-28 per cent reduction in emissions compared to 2005 levels by the year 2030. A six per cent reduction led by the people is at least a good starting point.

Once a government sees a groundswell of movement and opinions quickly changing, you can bet your voting opinion that they will be very quick to jump on the bandwagon. You might think that we have a poor policy in relation to greenhouse gas emission standards for cars sold in Australia. Not only is that not correct – there is NO standard in place at all. I can see this quickly changing once it is demanded by public action. Unfortunately, I no longer think it is correct to call our politicians our leaders. They are now followers. They follow public opinion so we have a huge opportunity to shape that opinion and demand policy that reflects our desire.

We don't necessarily have to like being in lockdown due to COVID-19, but most people are generally accepting of the science behind the decision to keep us apart. In much the same way, most people are now accepting of the science behind Climate Change – even if we don't like it. Unlike our current pandemic however there still seems to be very little action from government to protect the planet from ourselves. It is up to every single one of us to make small changes that collectively will make a difference.

The question is...are you prepared to make this compromise for the benefit of future you?

Mathew Dickerson is not a Climate Change Scientist. Mathew takes his car to a mechanic, visits his doctor when sick and uses an accountant to lodge his tax returns. He also has confidence in the scientific process. He is applying logic to a problem that is discussed in society every single day. PUBLISH DATE: 14 SEP 2020

About The Author

Mathew is rare in the technology journalist space. He doesn't just talk the talk – he walks the walk. His IT career started as a 13-year-old when his school made a progressive decision at the time and bought two devices that teachers had been talking about at conferences. Computers. Apple computers to be exact.

To say Mathew took to them like a duck to water is an understatement. Within days, Mathew was writing programs for them and quickly secured the job of entering the school student body into a database (possibly because most of the teachers didn't know how to turn on a computer). From this time on, Mathew realised that he had found his passion – technology.

Mathew started his first micro-business at the age of 12 and started his first IT company (AXXIS Technology) on 4 December 1989 as a brash and naïve 21-year-old. After the success of this business, Mathew went on to spin-off more technology start-ups. ComStall – a communication installation and cabling business. axxis ComWorld – a mobile phone retail business. 3ti - the technology training institute. Small Business Ru!es – an IT and business consultancy service. He also developed the SLAM DVD which helped IT businesses in 17 countries transform from a break/fix model to a Managed Services model. This was many years before the MSP concept became the only way to run an IT business.

These businesses collected a plethora of awards while under Mathew's guidance. The Microsoft Worldwide Partner of the Year award; the Microsoft Asia Pacific Certified Partner award; the Australian Business Council award for Innovation; the Australian Small Business Champions award for the best IT business in Australia; the Microsoft Australia/New Zealand Platform Partner of the Year; The Australian Small Business Champions award for the best regional business in NSW; the Telstra Partner award for Most Improved Independent Store in Australia; the MSPMentor USA award as a Top 100 Worldwide IT businesses and a Gold Rhino award as the best overall business in the region as well as multiple Silver Rhino awards for service; employment practices; presentation and marketing. Mathew's businesses have collected almost thirty major awards over the last two decades.

Mathew has also collected many personal accolades. He was showcased as the only Australian in the US published book *'Top 15 Successful SMB Consultants Worldwide'* and featured in the life advice book *'What I wish I knew at eighteen'*. He was named by MSPmentor as one of the Top 250 MSP individuals worldwide. He has won the Consensus IT Writers Award as the Most Entertaining IT Writer in Australia and a previous book of his won a bronze medal in the Axiom Worldwide book awards.

Mathew's business success gives him a 'hands on' perspective to take his knowledge to an international audience – which is exactly what he has been doing since 2004. That was the time when he first started speaking at conferences across the world for organisations such as Microsoft, Intel, Telecom, Kaseya and Autotask and has performed session and keynote presentations here in Australia as well as through locations such as Europe, New Zealand, South Africa, Sri Lanka, the United Arab Emirates and the United States.

Not only does Mathew perform presentations but talks and writes technology. He has a current radio segment called *'Tech Talk'* where he updates audiences on a weekly or fortnightly basis for nine different radio stations. His podcast, *'Tech Talk with Mathew Dickerson'* sits in the Top 10 locally produced technology podcasts across the nation. He writes a weekly column, again called *'Tech Talk'*, which appears in 140 mastheads across Australia and has written for a variety of publications such as Computer Reseller News with *'From the Coal Face'* and APC Magazine and TechGenix in the USA. Mathew is a highly respected IT entrepreneur and commentator in the Australian technology, IT and SMB space.

When he does find some spare time, he races Mountain Bikes with his local club; writes and recites poetry and spends time with his wife of twenty-six years and four children. He is also a dedicated community member having recently spent five years as Mayor of his city and now regularly organises fundraisers for important charities and facilities including securing several Guinness World Records as part of those events. He is an active member and current President of his local Rotary Club and enjoys his Mensa membership.

'Tech Talk 2021' is his fifth published work. You can contact Mathew at md@mathewdickerson.com

Mathew™ DICKERSON
TECHNOLOGIST · FUTURIST
SPEAKER · AUTHOR · POET

Made in United States
Troutdale, OR
11/03/2026

56844462R00106